Generative AI in Banking Financial Services and Insurance

A Guide to Use Cases, Approaches, and Insights

Anshul Saxena
Shalaka Verma
Jayant Mahajan

Apress®

Generative AI in Banking Financial Services and Insurance: A Guide to Use Cases, Approaches, and Insights

Anshul Saxena
Christ University, Lavasa
Maharashtra, India

Shalaka Verma
Microsoft, Mumbai
Maharashtra, India

Jayant Mahajan
Lavasa, Maharashtra, India

ISBN-13 (pbk): 979-8-8688-0558-5
https://doi.org/10.1007/979-8-8688-0559-2

ISBN-13 (electronic): 979-8-8688-0559-2

Copyright © 2024 by Anshul Saxena, Shalaka Verma, Jayant Mahajan

This work is subject to copyright. All rights are reserved by the Publisher, whether the whole or part of the material is concerned, specifically the rights of translation, reprinting, reuse of illustrations, recitation, broadcasting, reproduction on microfilms or in any other physical way, and transmission or information storage and retrieval, electronic adaptation, computer software, or by similar or dissimilar methodology now known or hereafter developed.

Trademarked names, logos, and images may appear in this book. Rather than use a trademark symbol with every occurrence of a trademarked name, logo, or image we use the names, logos, and images only in an editorial fashion and to the benefit of the trademark owner, with no intention of infringement of the trademark.

The use in this publication of trade names, trademarks, service marks, and similar terms, even if they are not identified as such, is not to be taken as an expression of opinion as to whether or not they are subject to proprietary rights.

While the advice and information in this book are believed to be true and accurate at the date of publication, neither the authors nor the editors nor the publisher can accept any legal responsibility for any errors or omissions that may be made. The publisher makes no warranty, express or implied, with respect to the material contained herein.

Managing Director, Apress Media LLC: Welmoed Spahr
Acquisitions Editor: Celestin Suresh John
Development Editor: Laura Berendson
Editorial Assistant: Gryffin Winkler

Cover designed by eStudioCalamar

Cover image designed by zaie on Freepik

Distributed to the book trade worldwide by Springer Science+Business Media New York, 1 New York Plaza, Suite 4600, New York, NY 10004-1562, USA. Phone 1-800-SPRINGER, fax (201) 348-4505, e-mail orders-ny@springer-sbm.com, or visit www.springeronline.com. Apress Media, LLC is a California LLC and the sole member (owner) is Springer Science + Business Media Finance Inc (SSBM Finance Inc). SSBM Finance Inc is a **Delaware** corporation.

For information on translations, please e-mail booktranslations@springernature.com; for reprint, paperback, or audio rights, please e-mail bookpermissions@springernature.com.

Apress titles may be purchased in bulk for academic, corporate, or promotional use. eBook versions and licenses are also available for most titles. For more information, reference our Print and eBook Bulk Sales web page at http://www.apress.com/bulk-sales.

Any source code or other supplementary material referenced by the author in this book is available to readers on GitHub. For more detailed information, please visit https://www.apress.com/gp/services/source-code.

If disposing of this product, please recycle the paper

Table of Contents

About the Authors vii

About the Technical Reviewer ix

Part I: Introduction to Generative AI 1

Chapter 1: Evolution of Generative AI 3
  1.1. Evolution of Generative AI 3
  1.2. Early Concepts and Theoretical Foundations of Generative AI 5
  1.3. Generative AI's Transformative Impact on Financial Verticals 9
  1.4. Roadmap for AI Implementation in BFSI 14
  1.5. Responsible AI 19
  1.6. Summary 23

Chapter 2: Technologies Behind Generative AI 25
  2.1. Historical Context and Foundations 25
  2.2. Key Models and Techniques 26
  2.3. Evaluation and Benchmarks 29
  2.4. Current Trends and Future Directions 31
  2.5. Future Directions and Research Opportunities 32
  2.6. Generative Adversarial Networks (GANs) 34
  2.7. Ethical and Privacy Challenges in Generative AI 36
  2.8. Regulatory Landscape and Policy Proposals 37

Chapter 3: Challenges and Potential Applications of Generative AI in BFSI 57
  3.1. Introduction 57
  3.2. Reimaging Banking Landscape with Generative AI Tools 63
  3.3. Current State of Financial Services and the Role of Technology 64
  3.4. Reimaging Landscape with Generative AI 70

TABLE OF CONTENTS

3.5. Current State of Insurance and the Role of Technology 72

3.6. Blueprint for Success: A Comprehensive Checklist for Enterprise Integration of Generative AI 77

3.7. Summary 80

Part II: Generative AI's Transformative Impact on Financial Verticals 83

Chapter 4: Transforming Banking: The Next Frontier 85

4.1. Introduction 85

4.2. Imperative Actions for Banking Modernization with Generative AI 88

4.3. Designing Banking Applications: Leveraging Generative AI for Innovation 99

4.4. Redefining Bank Business Support Functions with Generative AI 101

4.5. Designing Customer Support Chatbot 111

4.6. Summary 120

Chapter 5: Innovations in Investment Banking 123

5.1. Setting the Scene 123

5.2. AI-Driven Innovations in Finance 126

5.3. Preparing for a Revolutionized Future 128

5.4. The Transformative Potential of AI in Investment Banking and Trading 130

5.5. The New Era of Financial Planning and Advisory 136

5.6. Refined Portfolio Management Techniques 140

5.7. Challenges and Ethical Considerations 146

5.8. Challenges in AI-Driven Fraud Detection and Risk Management 155

5.9. Developing a Retrieval-Augmented Generation (RAG) Application for Stock Recommendations Using LlamaIndex 156

5.10. Summary 165

Chapter 6: Transformative Practices in Modern Financial Services 167

6.1. Introduction 167

6.2. Overview of the Traditional Financial Advisory Landscape 168

6.3. Impact on Long-Term Financial Planning 173

6.4. The New Era of Financial Planning and Advisory 175

6.5. Overview of the Current State of AI in Financial Services 178
6.6. Potential Disruptions Led by AI 181
6.7. AI's Role in Fostering Sustainable and Inclusive Finance 183
6.8. Ethical Considerations and Responsible AI Development 184
6.9. Role of Prompt Engineering 186
6.10. Summary... 199

Chapter 7: The Evolution of Insurance in the Digital Age 201

7.1. Introduction to Insurance Product Innovation 201
7.2. Enhancing Customer Engagement and Personalized Policies in the Insurance Industry with Generative AI 208
7.3. The Role of Technology in Customizing Policies 211
7.4. Improving Customer Loyalty and Satisfaction 213
7.5. Challenges in Personalization 214
7.6. Ethical Considerations and Pioneering Product Innovations in Insurance 218
7.7. Environmental and Climate Risk Insurance 220
7.8. The Impact of Digital Transformation on Compliance 231
7.9. Summary... 236

Part III: The Road Ahead for Generative AI in BFSI 239

Chapter 8: Roadmap for AI Implementation in BFSI 241

8.1. Assessment and Strategy Formation 242
8.2. Data Collection and Management 243
8.3. Integration and Scalability 247
8.4. Change Management for AI Adoption in BFSI 251
8.5. Continuous Monitoring and Feedback Loops in AI Systems for BFSI................... 257
8.6. Skill Development and Hiring... 260
8.7. Ethical and Regulatory Compliance in AI for BFSI.. 263
8.8. Summary.. 264

Chapter 9: Challenges in Mainstream Adoption 267
9.1. Challenges of Integrating AI into the BFSI Sector 267
9.2. Organizational Challenges 272
9.3. Training Needs for AI Integration in the Banking 280
9.4. Data Security Concern 288
9.5. Regulatory and Compliance Challenges 289
9.6. Summary 292

Chapter 10: Ethical Dilemmas and Future Potential of Generative AI in the Financial Sphere 295
10.1. Ethical and Responsible Concerns in Generative AI 296
10.2. Framing an AI Governance Policy for Generative AI in Banking 302
10.3. Cautionary Tales: Examples of What Could Go Wrong 308
10.4. The Transformative Potential of Generative AI in Banking 318
10.5. Embracing Digital Transformation in Banking 321
10.6. Summary 323

References 325

Appendix A: Glossary of Key Terms 327

Appendix B: List of Key Figures and Contributions 329

Index 331

About the Authors

Dr. Anshul Saxena is an author, corporate consultant, inventor, and educator who assists clients in finding financial solutions using quantum computing and Generative AI. He has filed three Indian patents and has been granted an Australian Innovation Patent. He has been instrumental in setting up new-age specializations such as decision sciences and business analytics in multiple business schools across India. Currently, he is working as Head – Center for Emerging Business Technologies at CHRIST University, Pune Lavasa Campus. Dr. Anshul has also worked with reputed companies such as IBM as a curriculum designer and trainer and has been instrumental in training 1,000+ working professionals from various corporate houses, including KPMG, IBM, Altran, TCS, METRO Cash & Carry, HPCL, and IOC. He has five years of experience in the domain of financial risk analytics with TCS and Northern Trust. Dr. Anshul holds a PhD in Applied AI (Management), an MBA in Finance, and a BSc in Chemistry. He possesses multiple certificates in Generative AI and quantum computing from organizations such as SAS, IBM, IISc, Harvard, and BIMTECH.

Ms. Shalaka Verma a passionate global technology leader with 23+ years of experience, currently leads the Azure Cloud Solution Architecture team at Microsoft India. Her core focus is ensuring net new workloads go-live on Azure swiftly and securely, maintaining optimal performance and reliability with the right deployment architectures.

ABOUT THE AUTHORS

Shalaka has strong technology R&D leadership. She has filed three US patents in blockchain, storage virtualization, and big data. In her earlier roles, she has pioneered futuristic application pilots for mobile value-added services and has also been awarded with a Gold Medal as a Homi Bhabha Award winner for her achievements as a computer scientist in BARC.

Dr. Jayant Mahajan is an Associate Professor in the Department of Management, distinguished by a diverse academic and personal profile. Holding a bachelor's degree in Physics and a master's degree in Information Management from the University of Mumbai, he further solidified his academic credentials with a PhD from Symbiosis International University. His doctoral research was instrumental in creating a training model for teaching ICT to visually challenged students. Beyond his academic pursuits, Dr. Mahajan's interests extend to strategy management, fintech, and the philosophy of technology, with a personal inclination toward triathlons and the nuances of metaphysics and cleantech.

About the Technical Reviewer

Dr. Abhijeet Birari is a passionate trader and trainer in the stock market for over a decade. He has a PhD in Derivatives Market and an MBA in Finance. He is an author, analyst, influencer, top B-School professor, and a coach to over 5,000 students in finance and stock market.

PART I

Introduction to Generative AI

CHAPTER 1

Evolution of Generative AI

1.1. Evolution of Generative AI

The evolution of Generative Artificial Intelligence (GenAI) represents a significant trajectory in the broader landscape of computational advancements, characterized by progressive shifts from rule-based systems to sophisticated neural networks and deep learning architectures. Originating in the 1950s, early efforts in AI focused on deterministic rule-based approaches, where systems were programmed to follow explicit instructions. These systems were limited by their rigidity and inability to adapt to new data. As a result, the 1970s and 1980s witnessed a pivotal transition towards machine learning, emphasizing data-driven models that could learn from inputs and mimic human cognitive functions. This period laid the groundwork for more advanced generative models, marking a departure from static algorithms to dynamic, learning-based approaches.

The resurgence of neural networks in the late 1980s, particularly with the introduction of the backpropagation algorithm, marked a critical juncture in AI research. Neural networks, which had been sidelined for their computational inefficiencies, were reintroduced with enhanced capabilities, leading to the development of foundational generative models. These models demonstrated the potential for AI systems to not only process information but also to generate new content, a capability that was further amplified in the 2000s with the advent of deep learning. The deep learning era, fueled by increased computational power and access to large datasets, facilitated significant breakthroughs in text and image generation, culminating in the creation of Generative Adversarial Networks (GANs) by Ian Goodfellow and his colleagues in 2014. GANs revolutionized the field by enabling the generation of highly realistic images, finding applications across various domains, including art, entertainment, and scientific research.

CHAPTER 1   EVOLUTION OF GENERATIVE AI

In the financial sector, the integration of generative AI is driving transformative changes across multiple verticals, from banking to insurance. AI-driven models are enhancing customer service through personalization, optimizing credit assessments, and streamlining regulatory compliance. In investment banking, algorithmic trading has been significantly augmented by AI's ability to analyze large datasets and identify market trends in real-time, leading to more efficient and precise trading strategies. The insurance industry is also witnessing a digital overhaul, with AI systems improving risk analysis, claims processing, and fraud detection. These developments underscore the potential of generative AI to enhance operational efficiency, accuracy, and customer experience in financial services. However, these advancements are accompanied by significant challenges, particularly concerning data privacy and the ethical implications of AI-driven decisions. The requirement for vast amounts of sensitive financial data to train AI models raises critical concerns about data security and the potential for algorithmic bias, which can lead to unfair or discriminatory outcomes.

Ethical considerations in generative AI have become a focal point of academic and industry discussions, particularly as the technology becomes more integrated into society. Researchers such as Timnit Gebru and Joy Buolamwini have highlighted the dangers of bias in AI models, particularly in areas such as facial recognition and credit scoring, where biased algorithms can perpetuate existing societal inequalities. Furthermore, the potential misuse of AI-generated content, such as deepfakes, presents significant risks to public trust and societal stability. These ethical dilemmas necessitate the development of robust frameworks for responsible AI development, emphasizing transparency, accountability, and fairness.

Looking forward, the future of generative AI is expected to be shaped by interdisciplinary influences, incorporating insights from neuroscience, psychology, and sociology. This convergence of fields will likely lead to more sophisticated models capable of understanding and generating human-like language, recognizing emotions, and interacting more naturally with users. The integration of generative AI with other emerging technologies, such as quantum computing and brain-computer interfaces, holds the promise of further breakthroughs in AI capabilities. However, these advancements must be carefully managed to ensure that the benefits of generative AI are realized without compromising ethical standards or societal values. As the technology continues to evolve, ongoing research and dialogue will be essential in navigating the complex landscape of generative AI and its impact on various sectors.

1.2. Early Concepts and Theoretical Foundations of Generative AI

Generative AI has its roots in the formative years of artificial intelligence. During this time, researchers conducted experiments with rule-based systems and endeavored to equip computers with the ability to engage in creative pursuits. The aim was to broaden the scope of computational capabilities beyond mere arithmetic, delving into the realm of art and design.

The evolution of Generative AI (Table 1-1), from the 1950s to the 2020s, has marked a transformative journey in artificial intelligence. Initially, rule-based systems dominated the field, focusing on deterministic tasks based on predefined instructions. The 1970s and 1980s saw a paradigm shift towards machine learning, emphasizing data-driven models that could adapt and learn, laying the groundwork for more advanced generative models. The resurgence of neural networks in the late 1980s, particularly through the backpropagation algorithm, further enhanced AI's capabilities, leading to early generative models.

Table 1-1. Evolution of Generative AI

Decade	Development Phase	Key Features and Advances
1950s–1960s	Early AI and Computational Creativity	– Focus on rule-based systems. – Initial use of computers for creating simple artistic patterns, highlighting the potential for creative processes.
1970s–1980s	Evolution of Machine Learning	– Development of more advanced machine learning algorithms. – Emphasis on mimicking human cognitive processes, including creativity and content generation.
Late 1980s–1990s	Resurgence of Neural Networks	– Rediscovery and improvement of neural networks, especially with backpropagation. – Creation of foundational generative models.

(*continued*)

Table 1-1. (*continued*)

Decade	Development Phase	Key Features and Advances
2000s	Deep Learning Breakthroughs	– Introduction of deep learning, fueled by better computational power and larger datasets. – Advancements in text generation and basic image creation. – Emergence of early Generative Adversarial Networks (GANs).
2010s	Emergence of Advanced Generative Models	– Introduction of GANs by Ian Goodfellow and team. – Significant progress in NLP with models like Transformer, GPT series, and BERT. – Quantum leap in the quality of generated images and videos.
Late 2010s–2020s	Proliferation and Ethical Considerations	– Mainstream adoption in art, music, literature, and science. – Heightened realism in AI-generated content. – Rising ethical discussions on authenticity, copyright, and misuse (e.g., deepfakes).
2020s	Integration, Accessibility, and New Frontiers	– Wider integration into technological platforms, enhancing accessibility. – Exploration of emotion-aware generation, multi-modal AI, and generation of novel scientific hypotheses or creative works.

The 2000s witnessed significant breakthroughs in deep learning, with advancements in text and image generation, culminating in the introduction of Generative Adversarial Networks (GANs) in 2014. GANs revolutionized image generation, finding applications across diverse fields. The 2010s brought about mainstream adoption and ethical considerations, highlighting the need for guidelines to govern AI-generated content. In the 2020s, generative AI has become increasingly integrated into various platforms, democratizing access and enabling innovative applications. Researchers continue to explore new frontiers, such as emotion-sensitive generation and multi-modal AI, pushing the boundaries of creativity and scientific discovery while addressing ethical implications in the ongoing evolution of AI.

Evolution of Large Language Models

The history of language intelligence within the field of artificial intelligence is a complex and multifaceted tale, comprising numerous developmental milestones, theoretical advancements, and technological breakthroughs. In order to examine this history in an organized manner (Table 1-2), it can be divided into several key areas of focus.

Table 1-2. *Evolution of LLM*

S.No.	Subtopic	Description
1	Early Foundations and Theoretical Concepts	Covers the initial ideas and theoretical underpinnings in linguistics, computer science, and cognitive science. Includes contributions of pioneers and the formulation of early concepts like machine translation and automated reasoning.
2	Machine Translation and Early Computational Linguistics	Focuses on early attempts at machine translation in the 1950s and 1960s, including the Georgetown experiment, and the development of computational linguistics as a field, emphasizing the analysis, processing, and simulation of human language.
3	The Rise of Rule-Based Systems	Characterized by the development of syntax-driven approaches and the use of formal grammars. Includes the creation of programming languages and frameworks for processing natural language.
4	AI Winter and Its Impact on Language Research	Explores the periods of reduced funding and interest in AI research, known as the "AI Winters," and their impact on the progress and direction of language-related AI research.
5	Statistical Methods and Corpus Linguistics	Describes the shift towards statistical methods in language processing due to more powerful computing resources and large language datasets. Highlights the significance of large corpora in language model development and training.

(*continued*)

Table 1-2. (*continued*)

S.No.	Subtopic	Description
6	The Emergence of Neural Networks and Deep Learning	Covers the resurgence of neural networks and the introduction of deep learning, detailing the development of models like LSTM networks and the transformer architecture and their impact on advancing language processing capabilities.
7	The Era of Large Language Models	Focuses on the development of large-scale, pre-trained language models such as GPT and BERT. This period is noted for significant advancements in natural language understanding and generation.

The development of language intelligence in artificial intelligence (AI) has been marked by significant milestones, each building upon interdisciplinary contributions from linguistics, computer science, and cognitive science. Early foundations were laid by pioneering figures like Noam Chomsky, whose generative grammar provided a theoretical basis for parsing and generating language computationally. The 1950s and 1960s saw the advent of machine translation, exemplified by the Georgetown experiment, which demonstrated the potential of computers to translate languages despite the limitations of early rule-based systems. The rise of computational linguistics and the integration of automated reasoning further advanced the field, setting the stage for more sophisticated language processing.

The late 1980s and 1990s introduced statistical methods and corpus linguistics, shifting from rule-based to data-driven approaches that leveraged large datasets and probabilistic models, such as Hidden Markov Models. The 21st century brought about a revolution with neural networks and deep learning, particularly through architectures like LSTM and the Transformer, which enhanced the ability of AI to process and generate human language. The era of large language models, typified by GPT and BERT, has since redefined language intelligence, pushing the boundaries of what AI can achieve in natural language processing.

1.3. Generative AI's Transformative Impact on Financial Verticals

Generative AI is rapidly transforming the financial ecosystem, fundamentally altering operational processes, customer experiences, and business strategies across various financial verticals. In the banking sector, Generative AI is revolutionizing core processes, including transactional activities, credit assessment, and regulatory compliance. AI-driven algorithms enhance customer support by providing personalized services and real-time assistance, while advanced credit scoring models offer more accurate risk assessments. Furthermore, regulatory compliance is becoming increasingly streamlined, with AI tools automating the monitoring and reporting of financial activities, ensuring adherence to complex legal frameworks.

In investment banking, the integration of AI is driving a digital transformation that is reshaping client interactions and algorithmic trading. AI-powered tools enable more efficient and precise trading strategies by analyzing vast datasets and identifying market trends in real time. Additionally, the use of AI in client management is enhancing the personalization of services, leading to improved client satisfaction and retention. Regulatory adaptations are also being facilitated by AI, allowing investment banks to navigate the ever-changing regulatory landscape more effectively.

The modern financial services landscape is similarly undergoing significant changes, with AI-driven techniques playing a pivotal role in portfolio management and financial planning. AI algorithms are optimizing portfolio management by assessing risk and predicting market movements, enabling more informed investment decisions. Moreover, AI is driving the development of customer-centric solutions, providing personalized financial advice, and creating next-generation financial planning tools that cater to the specific needs of individual clients. This shift towards personalization and efficiency is redefining the way financial services are delivered, making them more accessible and tailored to consumer demands.

In the insurance industry, Generative AI is leading a digital overhaul that is transforming risk analysis, claims processing, and fraud detection. AI-driven risk models are improving the accuracy of risk assessments, allowing insurers to offer more competitive pricing while minimizing potential losses. Claims processing is becoming more efficient with AI systems that can analyze claims data, detect anomalies, and expedite the approval process. Additionally, AI is playing a crucial role in fraud detection by identifying suspicious patterns and preventing fraudulent activities before they occur.

CHAPTER 1   EVOLUTION OF GENERATIVE AI

The innovation in product development is also noteworthy, with AI enabling insurers to design personalized insurance products that meet the evolving needs of consumers while ensuring compliance with regulatory requirements.

The integration of Generative AI across these financial verticals is not only enhancing operational efficiency but also driving innovation, personalization, and improved customer experiences. As the technology continues to evolve, its impact on the financial sector will likely expand, offering new opportunities and challenges for businesses and consumers alike. The intersection of human financial expertise with AI's predictive capabilities heralds a future where efficiency, accuracy, and innovation are paramount, reshaping the financial landscape in unprecedented ways.

Challenges and Potential Applications of Generative AI in BFSI

The integration of Generative AI within the Banking, Financial Services, and Insurance (BFSI) sector is both promising and complex, reflecting the evolving technological landscape of the industry. Currently, BFSI is deeply intertwined with advanced technologies such as machine learning, blockchain, and automation, which have revolutionized operations, customer service, and risk management. Generative AI, with its ability to create new data patterns, simulate financial scenarios, and enhance decision-making processes, holds significant potential to further transform the sector. However, its adoption is not without challenges. Data privacy remains a critical concern, as Generative AI requires vast amounts of sensitive financial data to function effectively. Ensuring the accuracy of AI models is another significant challenge, as errors in predictions or simulations could lead to substantial financial losses and erode trust in AI-driven solutions.

Moreover, the ethical implications of AI-generated content, particularly in areas like automated financial advice and fraud detection, necessitate careful consideration. Despite these challenges, the potential applications of Generative AI in BFSI are vast. It could revolutionize areas such as personalized financial products, automated risk assessments, and real-time fraud detection. As a precursor to a more detailed exploration, this chapter highlights the transformative impact Generative AI could have on BFSI, while also acknowledging the hurdles that must be overcome to realize its full potential. By examining these aspects, the chapter sets the stage for a comprehensive discussion on how BFSI can navigate the balance between innovation and responsibility in deploying Generative AI.

Transforming Banking: The Next Frontier

Generative AI is profoundly reshaping the Banking, Financial Services, and Insurance (BFSI) sector, fundamentally altering how key areas operate. Transactional operations are being revolutionized through the automation of routine processes, which enhances efficiency and minimizes human error. In the realm of customer engagement, AI-driven personalization is creating more tailored experiences, leading to increased customer satisfaction and loyalty. The credit process is also undergoing significant transformation, as AI models facilitate more accurate risk assessments and quicker decision-making, improving credit availability while reducing the likelihood of defaults. Security within the BFSI sector is being strengthened by AI's capacity to detect and respond to fraudulent activities in real time, offering a more proactive approach to risk management.

In the investment sphere, Generative AI is driving innovations in algorithmic trading, enabling more sophisticated strategies and optimizing portfolio management. Additionally, AI's role in regulatory compliance is becoming increasingly crucial, automating the monitoring and reporting of financial activities to ensure adherence to complex regulatory frameworks. These advancements, while promising, also introduce challenges such as concerns over data privacy, model accuracy, and the ethical implications of AI-driven decisions. Despite these hurdles, the integration of Generative AI in BFSI holds the potential for vast benefits, pushing the sector towards unprecedented levels of efficiency, innovation, and customer-centricity. As the technology continues to evolve, it is set to drive significant advancements while necessitating careful management of the risks and challenges associated with its deployment.

Innovations in Investment Banking

Generative AI is driving a profound transformation in investment banking, fundamentally altering traditional banking functions and paving the way for a digitally informed future. Algorithmic trading, a cornerstone of modern investment banking, is being significantly enhanced by AI's ability to analyze vast datasets and execute trades with precision and speed. These advancements not only optimize trading strategies but also increase market efficiency and reduce transaction costs. In portfolio management, Generative AI is enabling more sophisticated asset allocation and risk management techniques, allowing for real-time adjustments to portfolios based on dynamic market conditions and predictive analytics.

The role of investment advisors is also evolving, as AI tools augment their capabilities by providing data-driven insights and personalized recommendations. This shift enhances the advisor-client relationship by allowing for more tailored and strategic investment advice, thereby improving client outcomes. Additionally, Generative AI is playing a crucial role in forward-looking compliance measures. As regulatory environments become increasingly complex, AI-driven solutions are automating compliance monitoring and reporting, ensuring that investment banks adhere to regulations more efficiently and accurately.

However, the integration of Generative AI into investment banking is not without challenges. Concerns regarding data privacy, model transparency, and the ethical implications of AI-driven decision-making are significant considerations that must be addressed. Moreover, the reliance on AI systems necessitates robust cybersecurity measures to protect sensitive financial data. Despite these challenges, the trajectory of Generative AI in investment banking is undeniably promising. As the technology continues to evolve, it is set to drive further innovations, enhancing the efficiency, accuracy, and customer-centricity of investment banking services. The digital evolution of investment banking, informed by AI, represents a significant leap forward, promising a future where technology and human expertise work in tandem to navigate the complexities of global financial markets.

Transformative Practices in Modern Financial Services

The modern transformation of financial services is marked by significant advancements in portfolio management, trading predictions, and risk mitigation, driven largely by the integration of advanced technologies like Generative AI. Enhanced portfolio management now leverages AI-driven analytics to optimize asset allocation, adjust strategies in real time, and better predict market trends, thereby improving investment outcomes. Tailored financial solutions are increasingly becoming the norm, as AI enables more personalized services that align closely with individual client needs and preferences, fostering deeper client engagement and satisfaction.

In trading, innovative AI models are providing more accurate predictions, allowing financial institutions to anticipate market movements with greater precision. This not only improves trading efficiency but also enhances the ability to manage and mitigate risks. The evolution of financial planning and advisory services is equally significant. AI-driven tools are empowering advisors with data-rich insights, enabling them to offer more strategic and personalized advice. This redefinition of advisory services strengthens the client-advisor relationship by enhancing the relevance and accuracy of financial guidance.

Regulatory adherence is also undergoing a transformation, as AI automates compliance processes, ensuring that financial institutions meet increasingly complex regulatory requirements with greater efficiency and accuracy. However, these advancements are accompanied by challenges, including concerns over data privacy, the transparency of AI models, and the ethical implications of AI-driven decision-making. These issues must be carefully managed to ensure that the benefits of AI integration do not come at the cost of trust and security.

The future of financial services, as envisioned through the lens of Generative AI and other advanced technologies, is one of increased efficiency, personalization, and innovation. Yet, this future requires a balanced approach that addresses the inherent risks and ethical considerations while leveraging the transformative potential of technology to redefine the sector.

The Evolution of Insurance in the Digital Age

The insurance industry is undergoing a profound transformation, driven by advancements in technology that are reshaping risk assessment, customer engagement, and claims processing. Contemporary risk assessment techniques, powered by Generative AI and predictive analytics, enable insurers to evaluate risks with greater accuracy and granularity. This evolution enhances underwriting processes and allows for the development of more personalized insurance products that better align with individual customer needs. Enhanced customer engagement is also a critical area of innovation, as AI-driven platforms enable insurers to interact with clients in more meaningful and personalized ways, improving customer satisfaction and retention.

Advancements in claims processing have further streamlined the industry, with AI technologies automating the assessment and settlement of claims. This automation significantly reduces processing times, minimizes human error, and leads to quicker resolutions, ultimately enhancing the customer experience. Addressing fraud, a long-standing challenge in the industry, has also seen significant improvements. Machine learning algorithms are increasingly effective at detecting fraudulent patterns, enabling insurers to identify and mitigate fraud more efficiently.

The development of novel insurance products is another area where AI is making a substantial impact. Customized insurance offerings, designed to cater to specific customer segments and emerging risks, are becoming more prevalent. Regulatory compliance, traditionally a complex and time-consuming process, is also being revolutionized as AI tools help insurers navigate evolving regulatory landscapes more effectively. However, the integration of AI in insurance presents challenges, including concerns about data privacy, ethical implications of AI-driven decisions, and the potential for algorithmic bias.

As the insurance industry continues to evolve, these technological advancements are setting the stage for a future where insurance is more efficient, personalized, and responsive to the needs of both consumers and regulators. Balancing innovation with careful management of associated risks will be essential in shaping the industry's future trajectory.

1.4. Roadmap for AI Implementation in BFSI

Integrating AI into the Banking, Financial Services, and Insurance (BFSI) sector requires a structured and strategic approach, beginning with thorough initial assessments and progressing to the incorporation of expert insights. The process starts with a comprehensive evaluation of the organization's current technological infrastructure, identifying areas where AI can add the most value, such as risk management, customer service, or operational efficiency. This initial assessment should also include a cost-benefit analysis to ensure that the integration of AI aligns with the organization's strategic goals.

Following the assessment, it is essential to engage with industry experts to gain insights into the latest AI technologies and best practices. This step not only helps in selecting the most appropriate AI tools but also ensures that the implementation strategy is informed by cutting-edge developments and industry standards. Ethical considerations are paramount during this process, as AI in BFSI deals with sensitive financial data and decision-making processes. Establishing guidelines for data privacy, model transparency, and fairness is crucial to mitigate potential risks and build trust with stakeholders.

Practical steps in the integration process include pilot testing AI applications in controlled environments to identify potential challenges and refine the implementation strategy. Continuous training of staff is also necessary to ensure they are equipped to work alongside AI systems effectively. Moreover, the integration of AI should not be viewed as a one-time effort but as an ongoing process. Continuous adaptation to technological advancements is essential, requiring regular updates to AI systems and processes to maintain their relevance and effectiveness.

Incorporating AI into the BFSI sector presents both opportunities and challenges. While it can significantly enhance operational efficiency and customer experience, it also demands careful planning, ethical oversight, and a commitment to ongoing learning and adaptation. This structured approach ensures that AI integration is both strategic and sustainable, positioning the organization to thrive in an increasingly digital financial landscape.

Challenges in Mainstream Adoption

Integrating AI into the Banking, Financial Services, and Insurance (BFSI) sector presents a multifaceted set of challenges, encompassing technical, organizational, and regulatory obstacles. Technically, the complexity of AI systems, including the need for vast amounts of high-quality data, robust computational infrastructure, and advanced algorithmic models, poses significant hurdles. Organizations must also grapple with the challenge of integrating AI with existing legacy systems, which often lack the flexibility needed to accommodate modern AI technologies. Additionally, ensuring the accuracy and reliability of AI models is critical, as errors in financial predictions or risk assessments can lead to severe consequences.

Organizationally, the integration of AI requires a cultural shift, as employees must adapt to new technologies and workflows. Resistance to change, a lack of AI expertise, and the need for continuous training are common barriers. Effective change management strategies are essential to foster a culture of innovation and collaboration where AI is seen as a tool to enhance, rather than replace, human capabilities.

Regulatory challenges further complicate AI adoption in the BFSI sector. Financial institutions operate in highly regulated environments, and the deployment of AI must comply with stringent legal frameworks designed to protect consumer rights and ensure market stability. Regulatory uncertainty, particularly around the use of AI in decision-making processes, can create obstacles to adoption, as firms must navigate evolving guidelines and standards.

Market dynamics also play a critical role in AI adoption. Trust is a significant factor; consumers and stakeholders must have confidence in the fairness and transparency of AI-driven decisions. Furthermore, the competitive landscape influences AI integration, as firms seek to leverage AI to gain a competitive edge while managing the risks of over-reliance on technology.

Addressing these challenges requires a balanced approach that considers the technical, organizational, and regulatory complexities of AI integration. By understanding and mitigating these obstacles, the BFSI sector can harness the transformative potential of AI while maintaining trust and competitive advantage in a rapidly evolving market.

Ethical Dilemmas and Future Potential of Generative AI in the Financial Sphere

The ethical dimensions of AI in the Banking, Financial Services, and Insurance (BFSI) sector are critical, as the integration of advanced technologies raises significant challenges with far-reaching consequences. One of the primary ethical concerns revolves around data privacy, where AI's reliance on vast amounts of personal and financial data necessitates stringent measures to protect customer information. Additionally, the potential for algorithmic bias in AI systems can lead to unfair outcomes, such as discriminatory lending practices or inequitable risk assessments, which can disproportionately affect vulnerable populations.

Real-world consequences of unethical AI use in BFSI include erosion of trust, financial losses, and reputational damage. High-profile cases of biased algorithms or data breaches have underscored the need for a more responsible approach to AI implementation. In response, the tech industry has begun to address these issues by developing frameworks for ethical AI, such as principles for fairness, accountability, and transparency. However, the unique nature of the BFSI sector, with its focus on risk, regulation, and customer trust, requires more tailored approaches.

BFSI-specific nuances include the sector's regulatory environment, which demands compliance with strict legal standards while integrating AI technologies. Governance frameworks guiding AI's future trajectory in this context must therefore strike a balance between innovation and responsibility. These frameworks should ensure that AI applications are not only effective but also aligned with ethical standards that protect consumers and maintain market integrity.

Through case studies and industry insights, it becomes clear that a responsible approach to AI integration is essential for the BFSI sector. Such an approach involves continuous monitoring of AI systems, stakeholder engagement, and adherence to ethical principles that prioritize fairness and transparency. As AI continues to evolve, its successful integration in financial contexts will depend on the sector's ability to navigate these ethical dimensions thoughtfully and proactively, ensuring that technology serves the broader goals of trust, equity, and financial stability.

Challenges and Future Directions

Despite their remarkable capabilities, large language models also present significant challenges. One of the paramount concerns is the ethical ramifications of their use, including issues of bias, fairness, and privacy. These models can inadvertently perpetuate biases present in their training data, leading to skewed or unfair outcomes. Addressing these issues is a critical area of ongoing research and discourse. Another challenge is the environmental impact of training and operating such large models.

The computational resources required are substantial, raising questions about the sustainability of current approaches to artificial intelligence development.

Looking ahead, the field is poised for continued innovation. Research is ongoing into more efficient and ethical ways to train and deploy these models. There is also a growing interest in exploring how these models can be made more interpretable and transparent, which is essential for their responsible use.

The emergence of large language models has ushered in a new phase in the evolution of language intelligence. With models such as GPT and BERT at the forefront, this period has witnessed remarkable advancements in both natural language understanding and generation. As we navigate the challenges and opportunities presented by these technologies, their influence on the field of artificial intelligence and their broader societal implications will undoubtedly continue to be a subject of significant interest and importance.

Ethical and Societal Implications of Language Intelligence Technologies

The remarkable progress in language intelligence technologies has opened up a new era of opportunities and capabilities. However, this advancement also presents a range of ethical and societal implications that require careful deliberation. The incorporation of these technologies into various aspects of our lives has raised vital questions concerning privacy, bias, the potential for misuse, and the essential need for responsible development and deployment of AI in the domain of language technologies.

The rapid advancements in language intelligence technologies have ushered in an era of unprecedented opportunities, enhancing capabilities in communication, translation, and data analysis. However, these technological breakthroughs also bring forth a complex array of ethical and societal challenges that demand a thorough examination. One of the most pressing concerns is privacy, as the integration of language technologies into everyday life involves the collection and processing of vast amounts of personal data. This raises significant questions about how data is managed, who has access to it, and how it can be protected from misuse.

Bias in language models is another critical issue. AI systems trained on large datasets can inadvertently perpetuate or even amplify existing societal biases, leading to unfair or discriminatory outcomes. This concern extends beyond the technical domain, touching on deep-rooted issues of social justice and equity. The potential for misuse of these technologies further complicates the ethical landscape. For instance, AI-driven language tools could be exploited to create deepfake content, spread misinformation, or manipulate public opinion, thereby posing threats to societal stability and trust.

These challenges underscore the need for a responsible approach to the development and deployment of language intelligence technologies. Researchers and developers must prioritize fairness, transparency, and accountability in AI design, ensuring that these systems serve the public good. Furthermore, regulatory frameworks must evolve to address the unique risks posed by language technologies, balancing innovation with the protection of individual rights.

While the benefits of language intelligence technologies are immense, their integration into society must be guided by ethical principles that safeguard against potential harms. By adopting a multifaceted approach that considers technical, social, and ethical perspectives, stakeholders can ensure that the deployment of these technologies contributes positively to society, fostering trust and advancing human well-being.

Privacy Concerns

Undoubtedly, one of the most pressing ethical concerns in the realm of language intelligence systems is the issue of privacy. Given the necessity of large datasets for training these systems, the potential for sensitive personal information to be included is considerable. The use of such data, particularly without explicit consent, presents significant ethical challenges. For example, conversational AI systems, such as digital assistants, possess the capability to process and store personal and sensitive information discussed by users. It is of the utmost importance that these systems are designed and operated in a manner that respects user privacy and adheres to relevant data protection regulations.

Bias and Fairness

One of the significant challenges in the field of artificial intelligence is the potential for bias in AI language models. These biases often originate from the datasets used in training these models, which may reflect historical, cultural, or societal biases. The replication and amplification of these biases can result in unfair or discriminatory outcomes, particularly in applications such as recruitment, law enforcement, or credit scoring. To address these biases, it is essential to employ careful dataset curation, promote algorithmic transparency, and engage in continuous monitoring to guarantee fairness and equity in AI-driven decisions.

Potential for Misuse

The ethical implications of language intelligence technologies are a matter of great concern. The potential for these technologies to be used for malicious purposes, such as the creation of fake news or the automation of hate speech, is a serious issue that must be addressed. Furthermore, the ability of AI to generate realistic text poses a challenge in distinguishing between human and machine-generated content, which could exacerbate the problem of misinformation. It is essential that developers, policymakers, and users work together to ensure that these technologies are employed ethically and that their potential for misuse is mitigated.

1.5. Responsible AI

The need for responsible AI in the context of language technologies is paramount, as these systems increasingly influence various aspects of our lives. Ensuring transparency in how these technologies are built and operate is crucial for fostering trust and understanding among users. Transparency involves not only the clarity of algorithms and data usage but also making the decision-making processes of AI systems understandable to non-experts. This is particularly important in applications like automated content generation, language translation, and sentiment analysis, where the outcomes can significantly impact public discourse, individual privacy, and social dynamics. Moreover, accountability must be established in cases where AI systems produce errors or lead to harmful consequences. This means that developers and organizations deploying these technologies must be prepared to take responsibility for the actions of their AI, offering redress or correction when things go wrong. Without clear mechanisms for accountability, the deployment of AI in language technologies risks eroding public trust and could lead to unintended negative consequences.

Inclusivity in design and application is another critical aspect of responsible AI. Language technologies must be designed to accommodate diverse linguistic and cultural contexts, ensuring that they serve a broad spectrum of users. This requires careful consideration of the data on which these systems are trained, as well as the linguistic models that guide their outputs. Inclusivity also involves making these technologies accessible to people with disabilities, non-native speakers, and other marginalized groups, thereby preventing the digital divide from widening further. The importance of interdisciplinary collaboration cannot be overstated in this regard. Incorporating insights from fields like ethics, law, and social sciences is essential for guiding the development of AI technologies in a manner that aligns with societal values and norms. Ethical considerations must inform every stage of AI development, from the initial design and data collection phases to deployment and post-deployment monitoring.

The ethical and societal implications of language intelligence technologies present complex and multifaceted challenges. As these technologies continue to evolve and become deeply integrated into various societal domains, addressing these concerns becomes not just important, but imperative. The potential for language technologies to reshape communication, education, and even legal systems is immense, but with this potential comes the responsibility to ensure that these transformations are beneficial to society as a whole. This involves a concerted effort from multiple stakeholders—including AI researchers, developers, ethicists, policymakers, and users—to ensure that the advancement of language intelligence technologies is aligned with ethical principles and societal well-being.

Balancing innovation with responsibility is key to harnessing the benefits of AI in language processing while mitigating its risks. For instance, while AI-driven language models like GPT-3 have shown remarkable capabilities in generating human-like text, they also raise concerns about the propagation of biases, misinformation, and the potential misuse of these technologies in creating deepfakes or other forms of deceptive content. These risks underscore the need for robust ethical guidelines and governance frameworks that can guide the responsible development and use of language technologies. Policymakers must work closely with technologists to craft regulations that protect users while encouraging innovation. At the same time, continuous education and awareness-raising among the public are essential to ensure that users understand both the benefits and risks associated with these technologies.

One example of the challenges posed by AI in language technologies is the issue of bias in automated decision-making. Language models trained on large datasets can inadvertently learn and reinforce societal biases present in the data, leading to biased outcomes in areas such as hiring, law enforcement, and credit scoring. Addressing this requires not only technical solutions, such as developing algorithms that detect and mitigate bias, but also a broader commitment to ethical AI practices that prioritize fairness and equity. Another challenge is the protection of privacy in an era where language technologies can process and analyze vast amounts of personal data. Ensuring that AI systems are designed with privacy in mind and that they comply with relevant data protection laws is crucial for maintaining public trust.

The integration of AI into language technologies offers tremendous opportunities for enhancing human communication and understanding, but it also presents significant ethical and societal challenges that must be addressed. A responsible approach to AI development in this field involves ensuring transparency, accountability, and inclusivity while also fostering interdisciplinary collaboration to align these technologies with societal values. By balancing innovation with ethical considerations, stakeholders can ensure that the advancement of language intelligence technologies contributes positively to society, safeguarding against potential harms and maximizing their benefits. This balanced approach is essential for navigating the complex landscape of AI in language technologies, ensuring that their impact on society is both profound and positive.

Interdisciplinary Influences and Future Directions in Language Intelligence
The field of language intelligence, which lies at the intersection of artificial intelligence and linguistics, has been significantly impacted by insights from various disciplines, such as neuroscience, psychology, and sociology. This interdisciplinary convergence not only enhances the field but also paves the way for new directions and potential breakthroughs. As we anticipate the future, the incorporation of these diverse viewpoints is expected to play a critical role in shaping the development of language intelligence technologies.

Neuroscience: Understanding the Brain's Language Processing
The field of neuroscience provides profound insights into the intricate processes by which the human brain comprehends and produces language, which can inform the development of more sophisticated artificial intelligence (AI) models. By examining neural patterns and brain activity associated with language processing, researchers can gain valuable insights into the fundamental mechanisms of language handling. Such understanding may ultimately contribute to the development of AI systems that mimic the brain's efficiency and adaptability in language processing, potentially resulting in the creation of more natural and intuitive language AI systems.

Psychology: Cognitive and Behavioral Aspects
Psychology has a significant role in elucidating the cognitive and behavioral aspects of language utilization. Through the investigation of how individuals acquire, comprehend, and employ language, psychologists offer valuable insights into the nature of language acquisition and the cognitive processes that underlie comprehension and communication. This knowledge is vital for the development of artificial intelligence systems that can engage with humans in a more seamless and effective manner, particularly in areas such as education, therapy, and customer service.

Sociology: Language in Social Context

The discipline of sociology offers valuable perspectives on the operation of language within social contexts, including the norms, cultural subtleties, and societal trends that impact language use. To develop artificial intelligence systems that are sensitive to cultural nuances and responsive to social situations, it is crucial to comprehend these social aspects. This holds particular importance in areas such as social media monitoring, automated content curation, and multilingual communication platforms, where the social consequences of language are of paramount significance.

Future Directions and Potential Breakthroughs

In the realm of language intelligence, the horizon is filled with promising advancements. Among these are

The integration of language AI with Brain-Computer Interfaces (BCIs), which could provide groundbreaking communication solutions for individuals suffering from speech impairments. Additionally, this technology may facilitate novel modes of human-computer interaction, blurring the boundaries between thought and machine-mediated communication.

> **Emotional Intelligence in AI:** The development of AI systems capable of recognizing and responding to human emotions with increased accuracy is an area of significant potential. This could find applications in mental health, customer service, and entertainment.
>
> **Ethical and Fair AI:** As the field confronts ethical challenges, a greater emphasis on creating AI systems that are fair, unbiased, and transparent is anticipated. This involves not only technological advancements but also the establishment of regulatory frameworks and ethical guidelines for AI development and deployment.
>
> **Language Universality and Inclusivity:** Efforts to make AI more inclusive of diverse languages and dialects are crucial. Future research may concentrate on developing models that are proficient in non-dominant languages and cater to underrepresented linguistic communities.

AI and Creativity: Investigating the creative potential of AI in language, such as in literature, poetry, and scriptwriting, is an intriguing area for future exploration. This not only challenges the limits of AI's language capabilities but also raises philosophical questions about creativity and machine intelligence.

The future of language intelligence is inherently interdisciplinary, drawing from and contributing to a wide array of fields. As we venture into these uncharted territories, the fusion of diverse insights and perspectives will be instrumental in driving the field forward. The envisioned breakthroughs and innovations in language AI not only hold the promise of technological progress but also pose profound implications for our understanding of language, communication, and intelligence itself.

1.6. Summary

- Generative AI originated in the 1950s and 1960s, focusing on rule-based systems to enable computers to perform creative tasks, such as generating artistic patterns.

- The 1970s and 1980s marked a shift towards machine learning, moving from deterministic rule-based systems to data-driven models that could mimic human cognitive processes, including creativity.

- The late 1980s and 1990s saw a resurgence in neural networks, particularly through the backpropagation algorithm, laying the foundation for early generative models and advanced AI capabilities.

- The 2000s witnessed significant advancements in deep learning, leading to breakthroughs in text and image generation and the introduction of Generative Adversarial Networks (GANs) in 2014.

- The 2010s brought about mainstream adoption of generative AI, accompanied by increased discussions on ethical concerns, including authenticity, copyright, and the misuse of AI-generated content like deepfakes.

CHAPTER 1 EVOLUTION OF GENERATIVE AI

- In the 2020s, generative AI has become more integrated into various technological platforms, enhancing accessibility and leading to innovations such as emotion-sensitive generation and multi-modal AI.

- Generative AI is transforming the financial sector by revolutionizing processes in banking, investment banking, and insurance, improving operational efficiency, personalization, and regulatory compliance.

- The adoption of generative AI in the financial sector faces challenges such as data privacy, model accuracy, ethical implications, and integration with legacy systems, requiring careful management and strategic planning.

- The ethical dimensions of generative AI are critical, particularly concerning data privacy, algorithmic bias, and the potential misuse of AI-driven decisions, necessitating responsible AI development and deployment.

- The future of generative AI is shaped by interdisciplinary influences, with insights from neuroscience, psychology, and sociology contributing to advancements in AI's language processing, emotional intelligence, and ethical frameworks.

CHAPTER 2

Technologies Behind Generative AI

Introduction

Generative AI represents a transformative frontier in artificial intelligence, with the potential to revolutionize numerous fields, from natural language processing (NLP) and computer vision to music composition and drug discovery. At its core, generative AI focuses on creating new data instances that resemble a given dataset. This capability extends from generating realistic text and images to crafting intricate simulations and innovative designs. The architecture of generative AI encompasses a variety of models and techniques, each contributing uniquely to its evolution and application.

This chapter aims to provide an in-depth exploration of the architecture of generative AI. We will delve into historical developments, fundamental concepts, key advancements, and the intricate mechanisms that underpin these technologies. By examining these elements, we aim to offer a comprehensive understanding of how generative AI works, its potential applications, and the challenges it presents.

2.1. Historical Context and Foundations

Early Developments

The journey of generative AI began with early theoretical work on information theory and probability. Claude Shannon, often referred to as the father of information theory, made significant contributions that laid the groundwork for modern AI. His work on the mathematical theory of communication in the mid-20th century introduced concepts that would later become crucial in developing algorithms capable of generating data.

CHAPTER 2   TECHNOLOGIES BEHIND GENERATIVE AI

Markov Chains and Probabilistic Models

One of the earliest models used for generating sequences of data is the Markov chain, a mathematical system that transitions from one state to another within a finite set of states. These models are based on the principle that the future state depends only on the current state and not on the sequence of events that preceded it. This property, known as the Markov property, allows for the generation of sequences that statistically mirror the input data.

The Advent of Neural Networks

The introduction of neural networks brought a significant shift in the landscape of generative AI. Neural networks, inspired by the human brain's architecture, consist of interconnected nodes (neurons) that process data in layers. Early neural networks, such as the perceptron, laid the foundation for more complex architectures capable of learning from and generating data.

The Rise of Deep Learning

Deep learning, a subset of machine learning, involves neural networks with multiple layers (deep neural networks). This advancement allowed for more sophisticated data representation and learning. The ability of deep neural networks to model complex patterns in data made them particularly well-suited for generative tasks. Techniques such as convolutional neural networks (CNNs) and recurrent neural networks (RNNs) became instrumental in processing and generating images and sequences, respectively.

2.2. Key Models and Techniques

Autoencoders

Autoencoders are a type of neural network used to learn efficient representations of data. They consist of two main parts: an encoder that compresses the input data into a latent-space representation, and a decoder that reconstructs the original data from this representation. Variational Autoencoders (VAEs) introduced a probabilistic approach, allowing for the generation of new data points by sampling from the learned latent space.

Generative Adversarial Networks (GANs)

Generative Adversarial Networks, introduced by Ian Goodfellow and his colleagues in 2014, marked a significant breakthrough in generative modeling. GANs consist of two neural networks: a generator and a discriminator. The generator creates data instances, while the discriminator evaluates them against real data. Through an adversarial process, the generator improves its ability to create realistic data, leading to high-quality generative models.

The Transformer Era

Introduction to Transformers

The Transformer architecture, introduced in the paper "Attention is All You Need" by Vaswani et al. (2017), revolutionized NLP by addressing limitations in processing long-range dependencies in sequences. Transformers leverage an attention mechanism to weigh the influence of different input tokens, allowing for more effective context utilization. This architecture has become foundational in the development of generative AI models.

Evolution of Transformer Models

GPT Series

OpenAI's Generative Pre-trained Transformer (GPT) series exemplifies the power of transformers in generative AI. GPT-1 introduced the concept of pre-training on large text corpora followed by fine-tuning for specific tasks. GPT-2 demonstrated the effectiveness of scaling model size and training data, leading to impressive text generation capabilities. GPT-3 further scaled up the model, showcasing unprecedented language understanding and generation, and setting new benchmarks in AI performance.

BERT and Beyond

Bidirectional Encoder Representations from Transformers (BERT), developed by Google, brought a different approach by focusing on understanding context in both directions (bidirectional). While BERT is primarily used for understanding tasks, it has influenced generative models by improving contextual understanding. Other models like T5 (Text-To-Text Transfer Transformer) have built upon these ideas, providing versatile frameworks for generative and understanding tasks.

Applications and Impacts

The versatility of transformer models has enabled applications across various domains:

- **Text Generation**: From writing essays and poetry to generating code and dialogue, transformers have set new standards in text generation.

- **Translation**: Improved language translation models have emerged, leveraging the context-aware capabilities of transformers.

- **Summarization**: Generative models can create concise summaries of long documents, aiding in information retrieval and comprehension.

Reinforcement Learning from Human Feedback (RLHF)

Reinforcement Learning from Human Feedback (RLHF) is a technique where models learn from feedback provided by humans. This approach is particularly crucial for aligning AI behavior with human values and preferences, enhancing the utility and safety of generative models.

Preference modeling involves training a model to understand and predict human preferences. This can be achieved through supervised learning, where the model learns from labeled examples of preferred and non-preferred outputs. Reinforcement learning techniques, such as policy gradient methods, are then used to optimize the model's behavior based on these preferences.

Implementation Challenges

Implementing RLHF involves several challenges:

- **Data Collection**: Gathering high-quality, diverse, and representative feedback data is critical.

- **Scalability**: Scaling RLHF to large models and diverse tasks requires efficient algorithms and substantial computational resources.

- **Bias and Fairness**: Ensuring that the feedback does not introduce bias and that the model behaves fairly across different contexts is an ongoing concern.

Case Studies and Examples

ChatGPT

OpenAI's ChatGPT demonstrates the practical application of RLHF. By incorporating human feedback, ChatGPT improves its conversational abilities, making it more helpful and less likely to produce harmful outputs. The iterative process of collecting feedback, refining the model, and re-evaluating its performance showcases the effectiveness of RLHF.

Direct Preference Optimization (DPO)

Direct Preference Optimization (DPO) is an alternative approach to RLHF, focusing on direct gradient optimization based on preference data. Instead of learning a separate reward model, DPO optimizes the model's behavior directly using preference data. DPO simplifies the training process by eliminating the need for a separate reward model. This direct approach can be more efficient in terms of computational resources and implementation complexity.

Performance Gains

In certain scenarios, DPO has shown to provide performance gains comparable to or exceeding traditional RLHF methods. By directly optimizing for preferred outcomes, DPO can achieve alignment with human preferences more effectively.

Implementation and Results

Zephyr Model

The Zephyr model, developed by Hugging Face, exemplifies the success of DPO. By leveraging preference data and direct optimization techniques, Zephyr achieved significant improvements in performance and user satisfaction.

AI2's Efforts

The Allen Institute for AI (AI2) has also explored DPO, scaling it to large models with impressive results. Their work highlights the potential of DPO to enhance the capabilities and alignment of generative AI models.

2.3. Evaluation and Benchmarks

Importance of Evaluation

Evaluating generative models is crucial to understanding their performance, identifying areas for improvement, and ensuring they meet user expectations. Effective evaluation methods help in comparing different models and guiding further development:

> **Chatbot Arena:** Chatbot Arena provides a platform for human evaluators to compare responses from different chatbot models. While it offers valuable insights, the process can be slow and resource-intensive.
>
> **AlpacaEval:** AlpacaEval uses a set of predefined prompts and compares model responses against a base model. This automated approach provides quicker feedback, though it may face limitations in capturing the full complexity of human interactions.
>
> **MT-Bench:** MT-Bench scores model responses on a scale, providing a quantitative measure of performance. However, it can saturate with high-quality models, making it challenging to distinguish between top-performing models.

Open LLM Leaderboard: The Open LLM Leaderboard aggregates performance data across multiple models, offering a comprehensive overview. It serves as a valuable tool for tracking advancements and identifying leading models in the field.

WildBench: WildBench combines elements of chatbot evaluation and automated scoring, providing a faster and more comprehensive evaluation framework. It aims to balance the depth of human evaluation with the efficiency of automated methods.

The open-source community plays a vital role in advancing generative AI. The continuous release of new models contributes to the rapid evolution and democratization of AI technologies.

Alpaca, Vicuna, Koala, Dolly

These early open models demonstrated the feasibility of using synthetic and community-generated data for fine-tuning. Each model introduced unique techniques and datasets, contributing to the collective knowledge in the field.

Zephyr and Starling

Recent models like Zephyr and Starling have leveraged advanced techniques such as DPO and new preference datasets. Their success underscores the importance of innovative approaches in achieving state-of-the-art performance.

Community-Driven Projects

OpenAssistant

OpenAssistant represents a significant community effort to generate diverse and high-quality training data. By harnessing the power of crowdsourcing, OpenAssistant has created extensive datasets that are crucial for training and fine-tuning open models. This project illustrates the potential of community-driven initiatives to overcome data limitations and enhance the robustness and diversity of generative AI models.

Impact of Open Models on the Ecosystem

The proliferation of open models has democratized access to cutting-edge AI technologies, allowing researchers, developers, and enthusiasts to experiment and innovate without the need for extensive resources. This openness fosters collaboration, accelerates progress, and ensures that advancements in AI are shared broadly rather than being confined to a few proprietary entities.

2.4. Current Trends and Future Directions

Leveraging Synthetic Data

Synthetic data generation, such as CosmoPedia, has emerged as a vital tool for expanding training datasets. By generating realistic and diverse synthetic data, researchers can address data scarcity and improve the performance of generative models. Techniques like data augmentation and adversarial data generation are increasingly being used to enhance model robustness and generalization.

Exploring New Optimization Methods

The debate between PO and DPO continues to drive innovation in optimization methods. While DPO offers simplicity and efficiency, PO has shown potential advantages in certain scenarios. Researchers are actively exploring various optimization strategies to identify the most effective approaches for different tasks and model architectures.

Personalized Language Models

Personalization involves tailoring generative models to individual user preferences and contexts. Techniques such as user-specific fine-tuning and context-aware generation are being developed to enhance the relevance and utility of AI systems. Personalized models can provide more accurate and satisfying interactions, improving user experience across applications.

Developing Specific Evaluation Frameworks

Effective evaluation of personalized models requires new frameworks that account for individual preferences and contexts. Researchers are working on developing evaluation metrics and benchmarks that better reflect the nuances of personalized interactions. These frameworks aim to provide more meaningful insights into model performance and guide further refinement.

Ensuring Fairness and Mitigating Bias

One of the primary challenges in generative AI is ensuring fairness and mitigating bias in model outputs. This involves identifying and addressing biases in training data, developing fair evaluation metrics, and implementing techniques to ensure that models behave equitably across different user groups and contexts.

Enhancing Model Safety

Ensuring the safety of generative models involves preventing harmful outputs and aligning model behavior with ethical standards. Techniques such as reinforcement learning from human feedback (RLHF) and adversarial training are being used to enhance model safety. Ongoing research focuses on developing more effective methods for detecting and mitigating potential harms.

2.5. Future Directions and Research Opportunities

Multimodal Generative Models

Multimodal generative models, which can process and generate data across multiple modalities (e.g., text, image, audio), represent a promising direction for future research. These models have the potential to create more coherent and contextually rich outputs, enabling applications in areas such as multimedia content creation and cross-modal understanding.

Model Interpretability

Improving the interpretability of generative models is crucial for building trust and ensuring accountability. Researchers are developing techniques to make model decisions more transparent and understandable, enabling users to better grasp how and why certain outputs are generated. This transparency is essential for ethical deployment and governance of AI systems.

Collaborative and Decentralized AI

Collaborative and decentralized approaches to AI development, such as federated learning, are gaining traction. These approaches enable multiple parties to collaboratively train and fine-tune models without sharing raw data, preserving privacy and enhancing security. Decentralized AI development can foster greater inclusivity and democratization of AI technologies.

ChatGPT: A Practical Application of RLHF

OpenAI's ChatGPT serves as a practical example of the successful application of RLHF. By incorporating human feedback, ChatGPT has improved its conversational abilities, becoming more helpful and reducing the likelihood of harmful outputs. The iterative process of feedback collection, model refinement, and re-evaluation has been crucial in achieving these improvements.

Zephyr Model: Success with DPO

The Zephyr model developed by Hugging Face illustrates the potential of DPO. By leveraging preference data and direct optimization techniques, Zephyr achieved significant performance improvements. The model's success highlights the efficiency and effectiveness of DPO in aligning model behavior with user preferences.

AI2's Contributions to Scaling DPO

The Allen Institute for AI (AI2) has made notable contributions to scaling DPO to large models. Their work demonstrates that DPO can enhance the capabilities and alignment of generative models, providing valuable insights for future research and development.

AI2's efforts underscore the importance of scaling advanced optimization techniques to achieve state-of-the-art performance.

The architecture of generative AI is a dynamic and rapidly evolving field, driven by continuous advancements in model design, optimization techniques, and evaluation frameworks. From the foundational work of early pioneers to the transformative impact of modern architectures like transformers, generative AI has made significant strides in recent years.

The emergence of techniques like RLHF and DPO has further enhanced the alignment of generative models with human preferences, improving their utility and safety. The open-source community and collaborative projects have played a vital role in democratizing access to cutting-edge AI technologies, fostering innovation, and accelerating progress.

As we look to the future, addressing ethical and safety considerations, improving model interpretability, and exploring new research directions such as multimodal models and decentralized AI will be crucial. By continuing to push the boundaries of what is possible with generative AI, we can unlock new opportunities and applications that benefit society as a whole.

The journey of generative AI is far from over, and the ongoing research and development in this field promise to deliver even more exciting advancements in the years to come. Whether through enhanced personalization, improved safety, or novel applications, generative AI has the potential to transform our world in profound and positive ways.

Autoencoders and Variational Autoencoders (VAEs)

Autoencoders have been fundamental in learning data representations, especially for dimensionality reduction and feature learning. These neural networks consist of two main components: an encoder that compresses the input into a latent representation, and a decoder that reconstructs the input from this representation.

Standard Autoencoders

In a standard autoencoder, the encoder maps the input data to a latent space of lower dimensionality. The decoder then attempts to reconstruct the original data from this compressed representation. The model is trained by minimizing the reconstruction error, which measures the difference between the original input and its reconstruction. This process helps the model learn important features of the data.

Variational Autoencoders (VAEs)

VAEs, introduced by Kingma and Welling in 2013, add a probabilistic twist to the autoencoder architecture. Instead of mapping inputs to fixed points in the latent space, VAEs map them to distributions. The encoder outputs the parameters of a probability distribution (usually a Gaussian), and the decoder samples from this distribution to reconstruct the input.

The training objective of a VAE includes two parts: the reconstruction loss (similar to standard autoencoders) and the Kullback-Leibler (KL) divergence, which regularizes the latent space to follow a prior distribution (typically a standard normal distribution). This regularization ensures that the latent space is continuous and allows for smooth interpolation between data points, making VAEs particularly suitable for generative tasks.

2.6. Generative Adversarial Networks (GANs)

GANs, proposed by Goodfellow et al. in 2014, consist of two neural networks—the generator and the discriminator—that compete in a zero-sum game. The generator creates synthetic data samples, while the discriminator evaluates their authenticity compared to real data.

Architecture and Training Process

- **Generator**: Takes random noise as input and generates synthetic data samples.
- **Discriminator**: Receives both real and synthetic data samples and attempts to distinguish between them.

The training involves two simultaneous processes:

1. The generator tries to improve its output to make the synthetic data indistinguishable from real data.
2. The discriminator tries to get better at distinguishing real from synthetic data.

The generator's objective is to maximize the probability of the discriminator making a mistake, while the discriminator's objective is to minimize this probability. This adversarial process drives both networks to improve until the synthetic data is indistinguishable from the real data.

Applications and Variants of GANs

GANs have been used in various applications, including image generation, style transfer, and super-resolution. Several variants of GANs have been developed to address specific challenges and improve performance:

- **Conditional GANs (cGANs):** Introduce additional information (e.g., class labels) to both the generator and discriminator, allowing for conditional data generation.

- **CycleGAN:** Enables image-to-image translation without requiring paired training examples, useful for tasks like converting images from one domain to another (e.g., photos to paintings).

- **Wasserstein GANs (WGANs):** Modify the training objective to address instability in the training process, leading to more stable and reliable convergence.

Transformer Models and Their Impact

Transformers have revolutionized the field of NLP and beyond, providing a versatile architecture that can handle a wide range of tasks with remarkable efficiency and accuracy.

Attention Mechanism

The key innovation of transformers is the self-attention mechanism, which allows the model to weigh the importance of different parts of the input sequence when making predictions. This mechanism enables the model to capture long-range dependencies and contextual relationships more effectively than previous architectures like RNNs or LSTMs.

Transformer Architecture

A typical transformer consists of an encoder and a decoder, each composed of multiple layers of self-attention and feed-forward neural networks. The encoder processes the input sequence, while the decoder generates the output sequence based on the encoder's representations.

The scalability of transformers comes from their parallelizable architecture, which allows for efficient training on large datasets. This has enabled the development of massive pre-trained models that can be fine-tuned for specific tasks with relatively small amounts of task-specific data.

Impact of GPT Series

The GPT series (GPT-1, GPT-2, and GPT-3) has demonstrated the power of large-scale pre-training followed by fine-tuning. GPT-3, with 175 billion parameters, showcases unprecedented language understanding and generation capabilities, enabling applications such as text completion, translation, summarization, and even code generation.

Beyond NLP: Multimodal Transformers

Transformers have also been extended to handle multimodal data, such as combining text and images. Models like DALL-E and CLIP (Contrastive Language-Image Pre-training) leverage the transformer architecture to generate images from textual descriptions or understand images in the context of accompanying text. These advancements open up new possibilities for creative and practical applications in various domains.

2.7. Ethical and Privacy Challenges in Generative AI

Bias and Fairness

Generative AI models are trained on large datasets, which often contain biases present in the real world. These biases can be inadvertently learned and perpetuated by the models, leading to biased outputs that can reinforce stereotypes and unfair treatment.

Identifying and Mitigating Bias

Identifying bias involves analyzing the model's outputs across different demographic groups and contexts to detect disparities. Mitigating bias requires implementing techniques such as

> **Data Augmentation**: Adding diverse examples to the training data to ensure balanced representation
>
> **Bias Regularization**: Incorporating fairness constraints into the training process
>
> **Post-processing**: Adjusting the model's outputs to correct for identified biases

Ethical Considerations

Ethical considerations in generative AI extend beyond bias and fairness. They encompass issues such as transparency, accountability, and the potential for misuse. Developing ethical guidelines and frameworks for the responsible use of generative AI is crucial to ensuring that these technologies benefit society without causing harm.

Privacy Concerns

Generative AI models, especially those trained on sensitive data, raise significant privacy concerns. The ability of these models to generate realistic data increases the risk of privacy breaches and misuse.

Differential Privacy

Differential privacy is a technique that provides a formal privacy guarantee by adding noise to the training data or the model's outputs. This ensures that the inclusion or exclusion of any single data point does not significantly affect the model's predictions, thereby protecting individual privacy.

Federated Learning

Federated learning is a decentralized approach where models are trained locally on users' devices and only aggregated updates are shared with a central server. This method preserves data privacy by keeping the raw data on local devices and reducing the risk of exposure.

2.8. Regulatory Landscape and Policy Proposals

Existing Guidelines

Regulatory bodies and organizations have begun to establish guidelines and frameworks for the development and deployment of AI technologies. These guidelines aim to address issues such as data privacy, ethical use, and accountability.

General Data Protection Regulation (GDPR)

The GDPR, implemented by the European Union, sets strict requirements for data protection and privacy. It mandates transparency in data processing, user consent, and the right to access and delete personal data. Compliance with GDPR is essential for any AI system handling personal data.

AI Ethics Guidelines

Various organizations, including the IEEE and the Partnership on AI, have developed ethical guidelines for AI. These guidelines emphasize principles such as fairness, transparency, accountability, and the avoidance of harm.

Proposed Policies for Responsible Innovation

To foster responsible innovation in generative AI, several policy proposals have been put forward:

> **Algorithmic Accountability**: Requiring developers to conduct regular audits of their models to identify and mitigate biases and other ethical concerns
>
> **Transparency Mandates**: Ensuring that AI systems are transparent about how they make decisions and the data they use
>
> **Data Privacy Protections**: Strengthening data privacy regulations to safeguard user data and prevent misuse
>
> **Ethical AI Certification**: Establishing certification programs to verify that AI systems adhere to ethical guidelines and best practices

The Role of Stakeholders

Various stakeholders, including researchers, developers, policymakers, and users, play critical roles in fostering ethical use and continuous dialogue around generative AI:

> **Researchers and Developers**: Responsible for designing and implementing models that adhere to ethical standards and addressing any biases or fairness issues
>
> **Policymakers**: Tasked with creating and enforcing regulations that ensure the safe and ethical use of AI technologies
>
> **Users**: Need to be informed and empowered to understand how AI systems impact them and advocate for their rights

Multimodal and Interdisciplinary Research

Future advancements in generative AI will likely involve multimodal and interdisciplinary research. Combining insights from fields such as computer vision, linguistics, and cognitive science can lead to more robust and versatile models capable of handling complex tasks across various domains.

Enhancing Explainability and Trust

Improving the explainability of generative models is essential for building trust and ensuring accountability. Techniques such as attention visualization, model interpretability frameworks, and user-friendly explanations can help users understand how AI systems work and make decisions.

Collaborative Approaches to AI Development

Collaborative approaches, such as open-source development and participatory design, can drive innovation and ensure that generative AI technologies are developed with diverse perspectives and needs in mind. By involving a broad range of stakeholders in the development process, AI systems can be more inclusive and equitable.

Continuous Learning and Adaptation

Generative AI models must continuously learn and adapt to changing environments and user needs. Techniques such as online learning, active learning, and transfer learning enable models to stay up-to-date and improve over time. This continuous improvement is crucial for maintaining the relevance and effectiveness of generative AI systems.

The architecture of generative AI is a rich and evolving field, encompassing a wide range of models, techniques, and applications. From the foundational concepts of neural networks and probabilistic models to the transformative

impact of transformers and GANs, generative AI has made significant strides in recent years. This progress has opened up numerous possibilities, from enhancing natural language processing and computer vision to creating realistic simulations and innovative designs.

As we have seen, the journey of generative AI involves not only technical advancements but also ethical and societal considerations. Ensuring fairness, transparency, and accountability in AI systems is crucial for their responsible deployment. Privacy concerns, biases, and the potential for misuse must be addressed through rigorous evaluation, regulatory frameworks, and continuous dialogue among stakeholders.

Step-by-Step Process of Building LLMs

1. **Data Cleaning**

    Data cleaning is the first and one of the most critical steps in building LLMs. It involves preparing the raw text data to ensure it is free from errors, inconsistencies, and irrelevant information. The primary components of data cleaning are

    - **Data Filtering**: This process involves removing noise, handling outliers, addressing imbalances, and performing text preprocessing. Noise includes any irrelevant data that can hinder the model's learning process, while outliers are extreme values that can distort the model's performance. Addressing imbalances ensures that the dataset represents a diverse range of examples, preventing the model from becoming biased.

- **Deduplication**: Removing duplicate entries from the dataset is essential to avoid redundant information and to ensure that the model is exposed to a wide variety of unique data points.

By thoroughly cleaning the data, we create a solid foundation for the model, ensuring that it learns from high-quality, relevant information.

2. **Tokenization**

Tokenization is the process of breaking down text into smaller units called tokens, which can be individual words, subwords, or characters. This step is crucial for converting raw text into a format that can be processed by neural networks. Different tokenization methods include

- **Byte Pair Encoding (BPE)**: BPE is a subword tokenization technique that iteratively merges the most frequent pairs of bytes or characters to create a fixed-size vocabulary. It balances the granularity of tokens, capturing common subwords and entire words effectively.

- **Word Piece Encoding**: Similar to BPE, Word Piece Encoding splits words into subwords, optimizing the tokenization for handling large vocabularies and complex word forms. It is widely used in models like BERT (Bidirectional Encoder Representations from Transformers).

- **Sentence Piece Encoding**: This approach segments text into subwords or characters without requiring pre-tokenized inputs. It offers flexibility and efficiency, especially for languages with complex morphological structures.

Tokenization transforms text into sequences of tokens that can be efficiently processed by the subsequent layers of the model.

3. **Positional Encoding**

   Neural networks, particularly those based on the Transformer architecture, require a mechanism to understand the order of tokens within a sequence. Positional encoding provides this information, allowing the model to capture the relationships between tokens. Various methods of positional encoding include

   - **Absolute Positional Embeddings**: Assign fixed positional values to each token in the sequence. This method is straightforward but lacks flexibility in handling varying sequence lengths.

   - **Relative Positional Embeddings**: Encode the relative distances between tokens, allowing the model to adapt to different sequence lengths and structures.

   - **Rotary Position Embeddings**: Introduce rotational transformations to capture positional information, enhancing the model's ability to understand long-range dependencies.

   - **Relative Positional Bias**: Adjust attention weights based on the relative positions of tokens, improving the model's context awareness.

   Positional encoding is essential for models to maintain the sequential context of the input data, enabling more accurate and coherent text generation.

CHAPTER 2   TECHNOLOGIES BEHIND GENERATIVE AI

```
┌─────────────────────────────┐
│  Basic Building Blocks of LLMs │
└─────────────────────────────┘

     ( 1 )   Data Cleaning

     ( 2 )   Tokenization

     ( 3 )   Positional Encoding

     ( 4 )   LLM Architectures

     ( 5 )   Model Pre-training

     ( 6 )   Fine-Tuning
```

4. **LLM Architectures**

   The architecture of LLMs determines how the model processes and generates text. There are three main types of architectures:

   - **Encoder-Only**: Focuses on understanding the input data, making it suitable for tasks such as text classification and sentiment analysis. BERT is an example of an encoder-only model.

CHAPTER 2   TECHNOLOGIES BEHIND GENERATIVE AI

- **Decoder-Only**: Specializes in generating text based on the input context. These models, like GPT-3 (Generative Pre-trained Transformer 3), are used for text generation tasks.

- **Encoder-Decoder**: Combines both encoder and decoder components, making it ideal for tasks that require both understanding and generating text, such as language translation and summarization. The Transformer model introduced by Vaswani et al. is a prime example of an encoder-decoder architecture.

The choice of architecture depends on the specific application and the nature of the task at hand.

5. **Model Pre-training**

Pre-training is a fundamental phase where the model learns general language patterns from a large corpus of text. This phase involves several key techniques:

- **Masked Language Modeling (MLM)**: In MLM, certain words in a sentence are masked, and the model is trained to predict these masked words. This technique helps the model understand context and relationships between words. BERT uses MLM as one of its primary training objectives.

- **Causal Language Modeling**: In this approach, the model generates text by predicting the next word in a sequence based on the preceding words. This autoregressive method is used in models like GPT-3, where the goal is to generate coherent and contextually relevant text.

- **Next Sentence Prediction**: This technique trains the model to determine whether two sentences follow each other logically, improving the model's ability to understand discourse and generate coherent paragraphs. BERT also incorporates next sentence prediction in its pre-training.

- **Mixture of Experts**: This approach uses multiple specialized models (experts) and selects the appropriate one for each task, enhancing performance and efficiency. A mixture of experts can dynamically allocate computational resources, making the model more scalable.

Pre-training provides the model with a broad understanding of language, which can then be fine-tuned for specific tasks.

6. **Fine-Tuning and Instruction Tuning**

    After pre-training, the model undergoes fine-tuning to adapt to specific tasks or datasets. Fine-tuning involves adjusting the model's weights based on task-specific labeled data. There are several approaches to fine-tuning:

    - **Supervised Fine-Tuning**: Involves training the model on labeled data for specific tasks, such as question answering or sentiment analysis. This method ensures that the model performs well on targeted applications.

    - **General Fine-Tuning**: Applies general domain data to enhance the model's versatility, making it capable of handling a broader range of tasks.

    - **Multi-turn Instructions**: Adapts the model to handle multi-turn conversations, essential for developing dialogue systems and chatbots. This approach improves the model's ability to maintain context across multiple interactions.

    - **Instruction Following**: Trains the model to follow specific instructions, enhancing its ability to perform directed tasks and respond accurately to user commands.

Fine-tuning refines the model's performance, aligning it with the desired application and improving its accuracy and efficiency.

Challenges and Future Directions

Challenges in Building LLMs

Despite the advancements, building LLMs comes with several challenges:

- **Data Quality**: Ensuring high-quality, unbiased, and representative datasets is crucial for effective model training. Poor-quality data can lead to biased models and inaccurate outputs.

- **Computational Resources**: Training large models requires significant computational power and memory. Efficient use of resources and advancements in hardware are essential for scaling LLMs.

- **Ethical and Privacy Concerns**: Addressing biases, ensuring fairness, and protecting user privacy are critical issues. Transparent and accountable AI development practices are necessary to mitigate these concerns.

Future Directions

The field of LLMs is rapidly evolving, with several promising directions for future research and development:

- **Multimodal Models**: Integrating multiple data modalities, such as text, images, and audio, can enhance the model's capabilities and open up new applications.

- **Model Interpretability**: Improving the interpretability of LLMs is essential for building trust and ensuring accountability. Techniques that provide insights into the model's decision-making process can help users understand and validate its outputs.

- **Federated Learning**: This decentralized approach allows models to be trained across multiple devices without sharing raw data, preserving privacy and security. Federated learning can enable more collaborative and inclusive AI development.

- **Efficient Training Techniques**: Innovations in training techniques, such as sparsity and quantization, can reduce the computational requirements of LLMs, making them more accessible and scalable.

CHAPTER 2   TECHNOLOGIES BEHIND GENERATIVE AI

The architecture of Large Language Models is a complex and multifaceted domain, involving a series of carefully designed steps from data cleaning to fine-tuning. Each phase plays a crucial role in shaping the final model, ensuring it can understand and generate human-like text with high accuracy and efficiency. As LLMs continue to advance, addressing challenges such as data quality, computational resources, and ethical concerns will be paramount. The future holds exciting possibilities for LLMs, with advancements in multimodal integration, interpretability, and efficient training techniques paving the way for more powerful and versatile AI systems. By continuing to innovate and adhere to responsible AI development practices, we can unlock the full potential of LLMs to transform a wide range of applications and industries.

Introduction to AI-Powered Financial Planners

AI-powered financial planners utilize state-of-the-art technologies to provide personalized financial advice, portfolio management, risk assessment, and more. By incorporating advanced AI models, these systems can analyze vast amounts of data, recognize patterns, and offer insights that are both accurate and timely. This architecture combines various components, including data integration, machine learning models, and cloud platforms, to create a comprehensive financial planning solution.

Detailed Architecture for Financial Planners

1. **Cloud Platforms**

    Modern financial planners rely on robust cloud infrastructure to ensure scalability, reliability, and security. Major cloud providers such as AWS, GCP, Azure, and IBM offer services that support the deployment and management of AI models:

    - **AWS:** Provides tools like SageMaker for building, training, and deploying machine learning models at scale.

    - **GCP:** Offers AI Platform for managing machine learning workflows and BigQuery for data warehousing.

    - **Azure:** Azure Machine Learning enables data scientists to train and deploy models efficiently.

    - **IBM:** Watson Studio supports collaborative model development and deployment.

2. **Data Sources**

    A comprehensive financial planning system integrates diverse data sources to provide a holistic view of the financial landscape:

    - **News**: Financial news from sources like Bloomberg, Reuters, and The Wall Street Journal helps in understanding market trends.

    - **Social Media**: Platforms like Twitter, LinkedIn, and Reddit provide insights into public sentiment and trending topics.

    - **Financial Statements**: Official documents from SEC, NYSE, and other financial institutions offer authoritative data on companies.

    - **Market Data**: Real-time data from stock exchanges, forex markets, and commodity markets are crucial for timely decision-making.

    - **Economic Indicators**: Data from sources like the World Bank and IMF provide macroeconomic context.

    - **Proprietary Data**: Internal data such as client portfolios, transaction histories, and CRM data are used for personalized advice.

3. **Data Engineering**

    Data engineering processes ensure that raw data is transformed into a format suitable for analysis and model training:

    - **Data Cleaning**: Removes noise, corrects errors, and standardizes data formats to ensure quality and consistency.

    - **Tokenization**: Converts text data into tokens, enabling models to process and understand natural language inputs.

    - **Vector Embedding**: Transforms tokens into numerical vectors, capturing semantic meanings for further analysis.

    - **Feature Extraction**: Identifies and extracts key features relevant to financial planning tasks.

    - **Data Augmentation**: Enhances datasets by generating new examples through various augmentation techniques.

CHAPTER 2   TECHNOLOGIES BEHIND GENERATIVE AI

```
APIs
  ⑥ Robo-advisor    | Financial Education | ESG Scoring          | Fraud Detection
  ⑤ Portfolio Optimization | Retirement Planning | Insurance Planning | Estate Planning
  ④ Prompt Construction                    | LLM APIs
  ③ Named-Entity Recognition | Sentiment Analysis | Information Extraction | Terminology Understanding
  ② Data Cleaning   | Tokenization        | Vector Embedding     | Feature Extraction
  ① News            | Social Media        | Market Data          | Proprietary Data
                         Cloud Platforms
```

① Data Sources   ② Data Engineering   ③ Financial NLP   ④ LLMs (Large Language Models)   ⑤ Financial Planning Tasks   ⑥ Applications

4. **Data Integration**

   Effective data integration combines multiple data sources into a unified dataset, enabling comprehensive analysis and real-time decision-making:

   - **Real-time Data Pipeline APIs**: Streamline data flow from various sources into the system.
   - **Streaming Data**: Handles continuous data streams, ensuring that the system remains up-to-date with the latest information.
   - **Vector Database**: Stores vector embeddings and supports efficient retrieval and analysis of large datasets.

5. **Financial NLP (Natural Language Processing)**
   Financial NLP specializes in processing and understanding financial text data, providing capabilities such as

   - **Named-Entity Recognition (NER)**: Identifies and classifies entities such as companies, products, and events in financial texts.
   - **Sentiment Analysis**: Assesses the sentiment expressed in news articles, social media posts, and other textual data.

CHAPTER 2 TECHNOLOGIES BEHIND GENERATIVE AI

- **Information Extraction**: Extracts relevant information from unstructured text to populate databases and inform decision-making.

- **Terminology Understanding**: Recognizes and interprets specialized financial terminology.

6. **LLMs (Large Language Models)**

    LLMs form the backbone of advanced financial planning systems, offering powerful capabilities in natural language understanding and generation:

    - **Prompt Construction**: Creates prompts for LLMs to generate meaningful and contextually relevant responses. Techniques include:

    - **Retrieval-Augmented Generation (RAG)**: Enhances text generation by incorporating retrieved documents.

    - **Chain-of-Thought**: Guides the model through logical reasoning steps to arrive at accurate conclusions.

    - **LLM APIs**: Provides access to cutting-edge models like GPT-4, PaLM, and Claude for various NLP tasks.

7. **Financial Planning Tasks**

    AI models are trained to perform specific financial planning tasks, ensuring that the system can offer comprehensive support:

    - **Portfolio Optimization**: Balances risk and return to optimize investment portfolios.

    - **Risk Management**: Identifies, assesses, and mitigates financial risks.

    - **Retirement Planning**: Develops strategies to ensure financial security in retirement.

    - **Tax Optimization**: Identifies tax-saving opportunities and optimizes tax liabilities.

- **Debt Management**: Provides strategies for managing and reducing debt.
- **Insurance Planning**: Assesses insurance needs and recommends appropriate products.
- **Education Planning**: Develops savings plans for educational expenses.
- **Estate Planning**: Assists in managing and transferring assets in accordance with clients' wishes.

8. **Applications**

   Advanced AI enables a range of financial planning applications tailored to meet diverse client needs:

   - **Robo-advisor**: Provides automated investment advice based on user inputs and market conditions.
   - **Financial Sentiment Analysis**: Assesses overall market sentiment and predicts potential market movements.
   - **Quantitative Trading**: Develops and executes algorithmic trading strategies.
   - **ESG Scoring**: Evaluates companies based on environmental, social, and governance criteria.
   - **Fraud Detection**: Identifies suspicious activities and potential financial fraud.
   - **Credit Scoring**: Assesses the creditworthiness of individuals and businesses.
   - **Financial Education**: Offers resources and tools for improving financial literacy.
   - **Other Applications**: Support various other financial services tailored to specific user requirements.

9. **Trainable Models**

    The architecture supports the training and fine-tuning of models to enhance their performance on specific financial tasks:

    - **GPT-4, PaLM, Claude, LLaMA, etc.**: Advanced LLMs that can be fine-tuned for specific applications in financial planning

10. **Fine-Tuning Methods**

    Fine-tuning methods are crucial for adapting pre-trained models to specific financial contexts:

    - **Low-Rank Adaptation (LoRA), QLoRA**: Techniques that enable efficient fine-tuning by adjusting low-rank components of the model.
    - **Reinforcement Learning**: Uses feedback from the environment to optimize model performance in dynamic financial scenarios.

11. **Deployment Options**

    The system supports both cloud-native and on-premises deployments, offering flexibility and control:

    - **Cloud Native**: Utilizes cloud infrastructure for scalable and flexible deployment.
    - **On-Premises**: Supports deployment on local servers for enhanced security and data control.

Reimagining the architecture of financial planners with advanced AI technologies can transform the way financial services are delivered. By integrating diverse data sources, leveraging cutting-edge NLP and LLMs, and utilizing robust data engineering practices, these systems can provide personalized, accurate, and timely financial advice. This comprehensive approach ensures that financial planners are equipped to handle the complexities of modern financial markets, ultimately enhancing client satisfaction and optimizing financial outcomes.

CHAPTER 2   TECHNOLOGIES BEHIND GENERATIVE AI

Summary

1. **Historical Context and Foundations:**
   - Generative AI has evolved from early theoretical work on information theory and probability to sophisticated neural networks and deep learning models.
   - The introduction of autoencoders, GANs, and transformers has significantly advanced the capabilities of generative models.

2. **Transformer Models and Their Impact:**
   - The transformer architecture, particularly through models like GPT-3, has revolutionized NLP by leveraging the self-attention mechanism.
   - Transformers have extended beyond text to multimodal applications, integrating text, images, and other data types for more comprehensive generative capabilities.

3. **Alignment Techniques: RLHF and DPO:**
   - Reinforcement Learning from Human Feedback (RLHF) and Direct Preference Optimization (DPO) are critical techniques for aligning AI models with human values and preferences.
   - These techniques improve model utility, safety, and user satisfaction, demonstrating significant performance gains.

4. **Evaluation and Benchmarks:**
   - Effective evaluation methods are essential for understanding model performance and guiding further development.
   - Tools like Chatbot Arena, AlpacaEval, MT-Bench, and Open LLM Leaderboard provide valuable insights into model capabilities and areas for improvement.

CHAPTER 2   TECHNOLOGIES BEHIND GENERATIVE AI

5. **Open Models and Ecosystem Growth:**
   - The open-source community plays a vital role in advancing generative AI, contributing to rapid innovation and democratization of AI technologies.
   - Community-driven projects like OpenAssistant illustrate the potential of collaborative efforts in overcoming data limitations and enhancing model robustness.

6. **Ethical and Privacy Challenges:**
   - Addressing bias, fairness, and privacy concerns is crucial for the responsible deployment of generative AI.
   - Techniques like differential privacy and federated learning help protect user data while enabling effective model training.

7. **Regulatory Landscape and Policy Proposals:**
   - Regulatory frameworks and ethical guidelines are essential for ensuring the safe and responsible use of generative AI.
   - Proposals for algorithmic accountability, transparency mandates, and ethical AI certification aim to foster responsible innovation.

8. **Future-Oriented Solutions and Continuous Improvement:**
   - Advancements in multimodal and interdisciplinary research, model explainability, and collaborative approaches are crucial for the future of generative AI.
   - Continuous learning and adaptation ensure that generative AI models remain relevant and effective in dynamic environments.

The Path Forward

The future of generative AI holds immense promise, with ongoing research and development poised to unlock new opportunities and applications. By addressing ethical and privacy challenges, improving model interpretability, and fostering collaboration among stakeholders, we can ensure that generative AI technologies are developed and deployed responsibly.

CHAPTER 2   TECHNOLOGIES BEHIND GENERATIVE AI

As we continue to push the boundaries of what is possible with generative AI, it is essential to keep in mind the broader societal impacts and strive for inclusive and equitable outcomes. By doing so, we can harness the power of generative AI to drive innovation, enhance human capabilities, and create a better future for all.

Key Highlights

- Generative AI has its roots in early theoretical work on information theory and probability, with significant contributions from figures like Claude Shannon, laying the groundwork for modern AI.

- The development of Markov chains and probabilistic models was foundational, enabling the generation of sequences that statistically resemble input data, an early form of data generation.

- The advent of neural networks marked a major shift, with architectures like perceptrons introducing the concept of learning from data, setting the stage for more complex generative models.

- Deep learning, particularly through deep neural networks, advanced the field by enabling the modeling of complex patterns, which is crucial for generative tasks such as image and text generation.

- Autoencoders and their variant, Variational Autoencoders (VAEs), introduced the concept of learning efficient data representations, allowing for the generation of new data points by sampling from a learned latent space.

- Generative Adversarial Networks (GANs), introduced in 2014, revolutionized generative modeling by pitting two neural networks against each other in a zero-sum game, significantly improving the realism of generated data.

- The Transformer architecture, introduced in 2017, became a foundational model in NLP, enabling the development of powerful generative models like GPT by effectively processing long-range dependencies in sequences.

- The rise of models like GPT-3 showcased the power of scaling in generative AI, with larger models trained on vast datasets demonstrating unprecedented language generation capabilities.

- Reinforcement Learning from Human Feedback (RLHF) has become crucial in aligning AI models with human values, enhancing their utility and safety in practical applications.

- As generative AI evolves, it is addressing challenges such as bias, fairness, and privacy concerns with ongoing research focused on improving model interpretability and ethical deployment to ensure responsible innovation.

CHAPTER 3

Challenges and Potential Applications of Generative AI in BFSI

3.1. Introduction

The Banking, Financial Services, and Insurance (BFSI) sector has undergone a significant transformation in recent years, with technology playing a pivotal role. Advanced technologies like artificial intelligence (AI), machine learning (ML), and blockchain have revolutionized BFSI operations. AI and ML aid in risk assessment and fraud detection, while blockchain enhances transaction security and transparency. The rise of mobile banking and digital platforms has increased convenience for customers, leading to a surge in online banking activities and reducing reliance on physical branches. Additionally, the use of big data and analytics has provided deeper insights into customer behavior, aiding in customized product offerings.

The emergence of FinTech companies has been disruptive, introducing innovative financial solutions such as digital wallets and peer-to-peer lending. This has led to a blend of collaboration and competition between traditional BFSI institutions and FinTech firms. However, as digital solutions proliferate, so do cybersecurity risks, making the BFSI sector a prime target for cyberattacks. This necessitates robust cybersecurity measures and adherence to regulations like GDPR and CCPA for data protection.

Technological advancements have enabled BFSI entities to offer personalized services, enhancing customer engagement and satisfaction. The adoption of omnichannel strategies ensures a seamless experience across various touchpoints.

However, these institutions face challenges in keeping pace with rapid technological changes and integrating them into existing systems. Future trends like quantum computing and augmented reality are expected to further impact the BFSI landscape. There is also a growing focus on sustainable and ethical banking practices, influenced by technology-driven transparency and accountability.

The BFSI sector's future is intricately linked to its ability to adapt to technological advancements. Balancing technological integration with cybersecurity, regulatory compliance, and customer-centric approaches will be crucial for its continued evolution and role in the global economy.

Current State of the Banking and the Role of Technology

The banking sector has undergone a significant transformation due to the integration of technology, a process often referred to as digital transformation. This shift is driven by several key technological advancements.

Firstly, there's a clear move towards digital and mobile banking. Banks have been investing in online platforms and mobile apps, enabling customers to conduct various transactions remotely, such as opening accounts and transferring funds. This not only offers convenience to customers but also allows banks to reach a wider audience, including those in remote or underserved areas.

Artificial intelligence (AI) and machine learning (ML) are playing pivotal roles in transforming banking operations. AI is used for personalized customer services, like chatbots and virtual assistants, and in backend operations for tasks like risk assessment and fraud detection. Meanwhile, ML models analyze large amounts of data to uncover patterns and insights, which assist in decision-making and predictive analytics.

Blockchain technology, known for its use in cryptocurrencies like Bitcoin, is also gaining attention in the banking world. It's being explored for its potential in securing transactions and increasing transparency. The decentralized nature of blockchain could revolutionize data storage and sharing in the banking sector, potentially reducing fraud and operational costs. As banking becomes more digital, cybersecurity has become a crucial focus. Banks are investing in advanced cybersecurity measures to protect sensitive financial data from cyber threats. These measures include encryption, multifactor authentication, and constant monitoring of network activities.

Regulatory Technology (RegTech), a subset of FinTech, is another area of focus. It involves using technology to aid compliance with banking regulations. Banks are adopting RegTech solutions to automate compliance tasks, reduce risks, and efficiently manage regulatory reporting. Technology has also enabled banks to offer more personalized services. By analyzing customer data, banks can provide tailored financial products and advice. Improved customer experience through technology includes AI for personalized communication and digital platforms for self-service options.

Looking ahead, emerging technologies like quantum computing, augmented reality, and the Internet of Things (IoT) are expected to further impact the banking sector. These technologies could introduce new ways of interaction, enhance security protocols, and offer more efficient ways to process and analyze data.

Technology has become integral to the banking sector. It drives innovations that enhance customer experience, improve security, and increase operational efficiency. The ongoing evolution of technology in banking suggests a future where financial services are more accessible, secure, and tailored to individual needs.

Challenges in BFSI

The banking sector is currently navigating through a series of operational challenges that are crucial for its efficient and effective functioning (Table 3-1). One of the main challenges is digital transformation, where banks are working hard to upgrade their old systems to meet today's digital needs without interrupting customer service. Along with this comes the threat of cyberattacks as more banking activities move online. Banks have to keep updating their security measures to keep customer data and money safe.

Table 3-1. List of Challenges in Banking Industry

Challenge Categories	Description
Digital Transformation	Banks are upgrading old systems to meet modern digital requirements while ensuring continuous customer service.
Cybersecurity	As banking activities increasingly move online, banks must continually update security measures to protect customer data and funds.
Regulatory Compliance	Banks face the challenge of adhering to ever-changing and varied regulations across different regions, requiring constant vigilance and effort.
Risk Management	Managing various risks, such as credit risk, market volatility, and operational issues, necessitates robust and flexible strategies to adapt to market changes.

(continued)

Table 3-1. (*continued*)

Challenge Categories	Description
Cost Management	Banks need to reduce costs without compromising efficiency or customer service.
Technology Integration	Incorporating new technologies like AI, blockchain, and cloud computing into existing operations is complex and costly.
Customer Retention and Growth	In a competitive market, banks must continually improve customer service to retain existing customers and attract new ones.
Data Management	Effective management and utilization of vast quantities of data are crucial for making informed decisions.
Talent Acquisition and Retention	Finding and retaining skilled professionals in technology and data analysis is increasingly challenging.
Fraud Detection and Prevention	Banks are constantly challenged to enhance their methods of detecting and preventing increasingly sophisticated fraud to protect themselves and their customers.

Another big task for banks is to follow all the rules and regulations, which keep changing and vary from place to place. This requires a lot of effort and attention. Banks also have to be good at managing different kinds of risks, like the risk of someone not paying back a loan, market changes, or problems within their operations. This means creating strong plans that can adapt when things change in the market.

Banks also have to be careful with their money. They need to cut costs without affecting how well they work or how they serve their customers. Adding new technologies like AI, blockchain, and cloud computing into their current operations is also tricky and expensive. Keeping customers happy and coming back is also key, especially when there are so many banks to choose from. Banks need to keep improving how they serve their customers to keep them and attract new ones. They also have a lot of information that they need to manage well and use to make smart decisions.

As banks change, they need to find and keep the right people who know about technology and data analysis. This is becoming harder to do. Lastly, with fraud becoming more complex, banks are constantly challenged to improve how they detect and prevent fraud to protect both themselves and their customers.

Case Study 1

JPMorgan Chase & Co., as one of the foremost global banking institutions, faces a significant dilemma in balancing the rapid advancement in digital banking technology with the need to maintain robust security measures and regulatory compliance. This dilemma is underscored by a key business problem: how to seamlessly integrate innovative technologies without disrupting customer service, while simultaneously managing operational costs and staying ahead in a fiercely competitive market.

Challenge 1: Digital Transformation

The bank's aggressive pursuit of digital transformation is a response to this dilemma. While they strive to meet the evolving digital preferences of customers, they face the challenge of upgrading legacy systems without affecting customer service continuity.

Challenge 2: Cybersecurity Threats

Cybersecurity becomes a critical issue as more banking operations move online. JPMorgan Chase spends heavily on cybersecurity to protect against increasing online threats, facing the ongoing challenge of safeguarding customer data and financial assets in an ever-evolving digital threat landscape.

Challenge 3: Regulatory Compliance

Navigating the complex world of banking regulations poses another significant challenge. The bank must constantly adapt to changing and diverse regulations, a task that requires substantial resources and continuous vigilance.

Challenge 4: Risk Management

Effective risk management is vital, especially given the uncertainties in the global financial markets. The bank needs to manage credit, market, and operational risks without compromising on service quality or financial performance.

Challenge 5: Cost Management

JPMorgan Chase faces the ongoing task of managing costs effectively while ensuring that the quality of their operations and customer service is not compromised, a challenge intensified by the investments needed for digital transformation and cybersecurity.

Challenge 6: Technological Integration

Integrating cutting-edge technologies such as AI and cloud computing into existing operations presents both an opportunity and a challenge, as it involves significant investment and a need for specialized talent.

Challenge 7: Customer Experience and Retention

Enhancing customer experience and retention in a highly competitive market is crucial. The bank must innovate continually in service offerings to differentiate itself from competitors. This includes not only improving their digital platforms but also ensuring that these platforms are user-friendly and meet the varied needs of a diverse customer base. The challenge lies in providing a seamless, efficient, and personalized banking experience that can attract new customers and keep existing ones loyal.

Challenge 8: Data Management and Analytics

JPMorgan Chase has to manage and analyze vast amounts of data effectively. This data is crucial for understanding customer behaviors, market trends, and risk management. The challenge is to harness this data to make informed strategic decisions, tailor financial products, and enhance customer service, all while ensuring data privacy and security.

Challenge 9: Talent Acquisition and Retention

The rapidly changing technological landscape in banking requires JPMorgan Chase to continually acquire and retain skilled professionals, particularly in areas like IT, data analysis, and cybersecurity. The challenge is in competing with other tech-driven industries to attract top talent and then providing an environment that encourages their growth and retention.

Challenge 10: Fraud Detection and Prevention

With the sophistication of financial frauds increasing, JPMorgan Chase is continually challenged to improve its fraud detection and prevention mechanisms. This involves implementing state-of-the-art technologies and constantly updating their fraud detection algorithms to stay ahead of fraudsters, ensuring the security of customer transactions and bank operations.

JPMorgan Chase & Co.'s efforts to navigate these challenges reflect a dynamic and strategic approach to banking in the modern era. Their ability to balance the need for digital transformation with robust security and compliance, manage costs effectively, and continuously innovate in customer service and technology integration positions them as a leader in the industry. This case study demonstrates the complexities and opportunities within the banking sector, emphasizing the need for constant evolution and adaptation in this fast-paced, technology-driven world.

All these challenges mean that banks have to be strategic and adopt advanced ways of doing things to keep growing and stay secure.

3.2. Reimaging Banking Landscape with Generative AI Tools

Generative AI, with its advanced capabilities to create, simulate, and predict, can play a transformative role in addressing the key operational challenges in the financial services sector. Below is a reimagined approach to each challenge using Generative AI:

1. **Regulatory Compliance:** Generative AI can automatically generate compliance reports and audit trails by processing and interpreting complex regulatory texts. It helps in staying updated with regulatory changes globally, ensuring compliance, and reducing the risk of penalties.

2. **Cybersecurity Threats:** In cybersecurity, Generative AI can simulate various cyber-attack scenarios to strengthen the financial firm's defense mechanisms. By continuously learning from new threats, it enhances the predictive capabilities of security systems, offering proactive protection.

3. **Technological Integration:** Generative AI can assist in seamless integration of new technologies with existing systems. It can simulate the outcomes of integration, identifying potential issues and opportunities for optimization, thereby reducing integration time and costs.

4. **Data Management and Analytics:** For data management, Generative AI can organize and interpret large datasets, generating actionable insights. This capability not only improves decision-making but also enhances the efficiency and accuracy of data analytics processes.

5. **Risk Management:** In risk management, Generative AI can simulate various market conditions and risk scenarios. These simulations can inform more robust risk mitigation strategies, helping financial institutions prepare for diverse market dynamics.

6. **Customer Experience and Engagement:** Generative AI can personalize customer experiences by creating tailored financial products and services based on individual customer data. It can also generate interactive and engaging digital content to enhance customer interactions.

7. **Cost Management and Efficiency:** Generative AI can optimize operational processes, identifying areas where efficiency can be increased and costs reduced. By analyzing historical expenditure data, it can propose more efficient budget allocations and resource utilization.

8. **Talent Acquisition and Retention:** Generative AI can streamline the recruitment process by generating job descriptions, screening applications, and even conducting initial interviews. It can also identify skill gaps and recommend training programs, aiding in employee development and retention.

9. **Market Volatility and Economic Changes:** Generative AI can predict market trends and economic shifts by analyzing vast amounts of market data. This predictive analysis helps financial institutions adapt their strategies proactively to changing market conditions.

10. **Fraud Detection and Prevention:** Generative AI can enhance fraud detection by simulating fraudulent activities and training detection systems. It continuously evolves by learning from new patterns of fraud, thereby strengthening the financial institution's capacity to detect and prevent fraud.

In each of these applications, Generative AI not only addresses the specific challenge but also adds value by enhancing efficiency, accuracy, and predictive capabilities. It represents a significant step forward in the digital transformation of the financial services sector.

3.3. Current State of Financial Services and the Role of Technology

In today's world, the financial services sector is undergoing a major transformation, largely driven by technology. This sector, which includes banking, wealth management, mutual funds, stock market activities, tax and audit consulting, and portfolio management, is rapidly evolving to adapt to the digital era. The digital revolution has changed how these services are delivered. Traditional methods are giving way to digital platforms and mobile apps. For example, a person can now easily manage their stock market investments or check their mutual fund performances on their smartphones. This digital shift is not just convenient for tech-savvy customers but also brings financial services to those who were previously excluded due to geographic or economic barriers.

Artificial intelligence (AI) and machine learning (ML) are significantly impacting these services. AI helps in providing round-the-clock customer support through chatbots, while ML assists in making accurate risk assessments, crucial in fields like portfolio management and wealth management. These technologies are also enhancing fraud detection, a critical aspect in maintaining trust in financial transactions.

Blockchain technology and cryptocurrencies are also influencing the sector. Blockchain's secure and transparent nature is appealing for secure transaction processing, which is vital in areas like mutual funds and stock market operations. Cryptocurrencies, despite their controversies, are pushing financial institutions to explore new digital currency possibilities.

Cybersecurity is another top priority. With the increase in online financial activities, protecting sensitive data is crucial. Financial institutions invest heavily in advanced security measures like encryption and multi-factor authentication to safeguard client information in services like tax consulting and audit.

Regulatory Technology (RegTech) is gaining importance due to the complex and ever-changing financial regulations. RegTech helps automate compliance processes, making it easier for institutions to stay compliant without compromising operational efficiency, which is essential in services like wealth management and portfolio management.

Personalization of services has been greatly enhanced by technology. Financial institutions now use customer data to offer tailored advice and products, enhancing customer experience across various services, whether it's customizing a mutual fund portfolio or offering tailored wealth management solutions.

Looking ahead, the sector is expected to be further influenced by emerging technologies like the Internet of Things (IoT), quantum computing, and augmented reality. These technologies will continue to revolutionize customer experiences, introduce innovative financial products, and streamline operations.

In summary, technology is at the forefront of reshaping the financial services sector, impacting everything from banking to wealth management and from the stock market to tax consulting. Institutions that successfully integrate these technological advancements while ensuring security and compliance are likely to lead in this dynamic and rapidly evolving industry.

CHAPTER 3　CHALLENGES AND POTENTIAL APPLICATIONS OF GENERATIVE AI IN BFSI

Challenges in Financial Services

The financial services industry, which includes areas like banking, wealth management, mutual funds, the stock market, tax/audit consulting, and portfolio management, is currently navigating through a host of challenges in its integration with technology (Table 3-2). One of the biggest hurdles is digital transformation. Financial institutions are working hard to shift from traditional methods to modern digital platforms. This change is necessary to keep up with customer demands in services such as online banking and digital wealth management, but it's not easy to upgrade these complex systems without affecting customer service.

Table 3-2. List of Challenges Faced in Financial Services

Challenge Categories	Description	Examples in Financial Services
Digital Transformation	Upgrading from traditional systems to modern digital platforms without disrupting customer service.	Implementing online banking and digital wealth management.
Cybersecurity Risks	Protecting sensitive client data and financial assets from cyberattacks as financial activities increase online.	Securing online transactions and stock trading platforms.
Regulatory Compliance	Keeping up with changing and diverse regulations across different regions.	Ensuring compliance in tax consulting and portfolio management.
Technological Integration	Incorporating new technologies like AI, ML, and blockchain into existing systems.	Using AI for personalized investment advice and real-time data analysis in the stock market.
Data Management and Privacy	Handling and securing large volumes of client data for informed decision-making.	Managing client data in portfolio management and wealth management.
Customer Experience and Personalization	Offering tailored and efficient customer service in a competitive market.	Personalizing services in mutual funds and banking.

(*continued*)

Table 3-2. (*continued*)

Challenge Categories	Description	Examples in Financial Services
Fraud Detection and Prevention	Implementing advanced systems to protect against sophisticated financial frauds.	Protecting assets in online banking and stock market transactions.
Talent Acquisition and Retention	Attracting and retaining skilled professionals in tech and data analytics.	Hiring and keeping experts in technology and data analytics for various financial services.
Adapting to Emerging Technologies	Preparing for and adopting emerging technologies like IoT, quantum computing, and augmented reality.	Enhancing stock market analysis and portfolio management with new technologies.
Balancing Cost with Innovation	Investing in new technologies and cybersecurity measures while managing operational costs.	Innovating in mutual funds and tax consulting without significantly increasing customer costs.

Another major concern is cybersecurity. As more financial activities, like stock trading and online transactions, move online, the risk of cyberattacks increases. Banks and other financial institutions are spending a lot to protect their clients' sensitive data from these threats. Then there's the issue of regulatory compliance. The rules in the financial world are constantly changing and vary from region to region. Keeping up with these regulations is especially important in areas like tax consulting and portfolio management.

Integrating new technologies such as AI, ML, and blockchain into existing operations is also a challenge. These technologies can greatly improve services like personalized investment advice in wealth management or real-time data analysis in the stock market, but they require significant investment and expertise.

Managing vast amounts of data is another challenge. Financial services providers must ensure the security and privacy of client data, which is essential for making informed decisions in areas like portfolio management. Providing personalized and efficient customer service is also crucial. Financial institutions are using technology to tailor their services to individual needs, but this must be balanced with maintaining high-quality customer service.

CHAPTER 3 CHALLENGES AND POTENTIAL APPLICATIONS OF GENERATIVE AI IN BFSI

Fraud detection and prevention are more important than ever with the increase in online financial transactions. Financial services are using advanced technology to protect their clients' assets, but fraudsters are always finding new methods. Attracting and retaining skilled professionals in technology and data analytics is becoming increasingly difficult in this competitive sector.

Emerging technologies like IoT, quantum computing, and augmented reality are on the horizon, and financial institutions need to be ready to adopt these to stay ahead, especially in areas like stock market analysis and advanced portfolio management. Lastly, balancing the cost of these new technologies and innovations with operational costs is a delicate act. Financial services need to innovate without significantly increasing costs for their customers, particularly in areas like mutual funds and tax consulting.

The financial services sector is facing a range of challenges as it integrates technology into every aspect of its operations. These challenges require strategic planning and a focus on customer-centric services to ensure growth and resilience in this rapidly evolving, tech-driven environment.

Case Study 2

Goldman Sachs, a global powerhouse in investment banking, securities, and investment management, presents an instructive case study on handling operational dilemmas in today's complex financial market.

Dilemma 1: Regulatory Compliance vs. Market Agility

Goldman Sachs operates in a regulatory minefield, facing the dilemma of maintaining strict compliance while needing to remain agile in a dynamic market. The firm has robust legal and compliance units, utilizing cutting-edge technology and expert staff to monitor and adapt to regulatory changes globally, balancing legal adherence with market responsiveness.

Dilemma 2: Cybersecurity vs. Digital Innovation

As cybersecurity threats escalate, Goldman Sachs confronts the challenge of fortifying its digital defenses without stifling technological innovation. The firm has developed a comprehensive cybersecurity framework, employing state-of-the-art technologies and a specialized IT security team to protect digital assets and client data while continuing to invest in digital advancements.

Dilemma 3: Technological Integration vs. Operational Stability

Goldman Sachs faces the task of integrating disruptive technologies like AI and blockchain into its existing operations. The firm has been an early adopter of digital innovations, investing in fintech startups and employing blockchain for secure transactions, carefully balancing these integrations with the stability of their core operations.

Dilemma 4: Data Utilization vs. Privacy Concerns

Data management is crucial for Goldman Sachs, yet it must navigate the delicate balance between leveraging data for insights and respecting privacy concerns. The firm employs sophisticated data systems and big data analytics for informed decision-making while prioritizing client data privacy and security.

Dilemma 5: Comprehensive Risk Management vs. Competitive Performance

Risk management is a cornerstone of Goldman Sachs' strategy, yet it must manage risks without compromising competitive performance. The firm uses advanced tools for risk assessment, addressing market, credit, and operational risks while striving to maintain its edge in the financial market.

Dilemma 6: Enhancing Customer Experience vs. Operational Efficiency

In the competitive financial market, Goldman Sachs aims to enhance customer experience while maintaining operational efficiency. This includes offering personalized financial advisory services and leveraging digital platforms for customer interaction, balancing customer-centric approaches with efficient operations.

Dilemma 7: Cost Management vs. Quality Service Delivery

Goldman Sachs faces the challenge of managing costs effectively while ensuring high-quality service. The firm adopts cost-saving measures and technology-driven process optimization, striving to deliver premium services without unnecessary expenditure.

Dilemma 8: Talent Acquisition vs. Cost Management

Attracting and retaining top talent, especially in technology and financial analysis, is vital for Goldman Sachs. The firm offers competitive salaries and development opportunities, balancing the cost of such investments with the benefits of a skilled workforce.

Dilemma 9: Navigating Market Volatility vs. Stable Growth

Goldman Sachs actively adapts to market trends and economic changes, balancing the need to mitigate market volatility impacts with the goal of achieving stable growth for the firm and its clients.

Dilemma 10: Fraud Detection vs. Customer Convenience

The firm implements advanced fraud detection systems, including AI and machine learning, to monitor transactions. This proactive approach must be balanced with ensuring that fraud prevention measures do not impede customer convenience or transaction efficiency.

Goldman Sachs' handling of these operational dilemmas in the financial services sector showcases a strategic balance between regulatory compliance, risk management, technological innovation, and maintaining market competitiveness. Their approach exemplifies the importance of agility and forward-thinking in navigating the complexities of today's financial landscape.

3.4. Reimaging Landscape with Generative AI

Generative AI, with its advanced capabilities to create, simulate, and predict, can play a transformative role in addressing the key operational challenges in the financial services sector. Below is a reimagined approach to each challenge using Generative AI:

1. **Regulatory Compliance:** Generative AI can automatically generate compliance reports and audit trails by processing and interpreting complex regulatory texts. It helps in staying updated with regulatory changes globally, ensuring compliance, and reducing the risk of penalties.

2. **Cybersecurity Threats:** In cybersecurity, Generative AI can simulate various cyber-attack scenarios to strengthen the financial firm's defense mechanisms. By continuously learning from new threats, it enhances the predictive capabilities of security systems, offering proactive protection.

3. **Technological Integration:** Generative AI can assist in seamless integration of new technologies with existing systems. It can simulate the outcomes of integration, identifying potential issues and opportunities for optimization, thereby reducing integration time and costs.

4. **Data Management and Analytics:** For data management, Generative AI can organize and interpret large datasets, generating actionable insights. This capability not only improves decision-making but also enhances the efficiency and accuracy of data analytics processes.

5. **Risk Management:** In risk management, Generative AI can simulate various market conditions and risk scenarios. These simulations can inform more robust risk mitigation strategies, helping financial institutions prepare for diverse market dynamics.

6. **Customer Experience and Engagement:** Generative AI can personalize customer experiences by creating tailored financial products and services based on individual customer data. It can also generate interactive and engaging digital content to enhance customer interactions.

7. **Cost Management and Efficiency:** Generative AI can optimize operational processes, identifying areas where efficiency can be increased and costs reduced. By analyzing historical expenditure data, it can propose more efficient budget allocations and resource utilization.

8. **Talent Acquisition and Retention:** Generative AI can streamline the recruitment process by generating job descriptions, screening applications, and even conducting initial interviews. It can also identify skill gaps and recommend training programs, aiding in employee development and retention.

9. **Market Volatility and Economic Changes:** Generative AI can predict market trends and economic shifts by analyzing vast amounts of market data. This predictive analysis helps financial institutions adapt their strategies proactively to changing market conditions.

10. **Fraud Detection and Prevention:** Generative AI can enhance fraud detection by simulating fraudulent activities and training detection systems. It continuously evolves by learning from new patterns of fraud, thereby strengthening the financial institution's capacity to detect and prevent fraud.

In each of these applications, Generative AI not only addresses the specific challenge but also adds value by enhancing efficiency, accuracy, and predictive capabilities. It represents a significant step forward in the digital transformation of the financial services sector.

3.5. Current State of Insurance and the Role of Technology

The insurance industry, traditionally seen as cautious and resistant to change, is currently experiencing a significant transformation driven by technology. This industry is moving away from its old paper-based methods to digital platforms. Now, customers can manage their policies and file claims online, making everything more efficient and convenient.

Artificial intelligence (AI) and predictive analytics are really changing the game in insurance. AI is being used to automate tasks like underwriting, making risk assessments more accurate, and tailoring policies to individual needs. Predictive analytics helps insurers look at big sets of data to spot trends and predict risks, improving decision-making in underwriting. Blockchain technology is another area that's being explored in insurance. It offers a way to keep records that are secure and decentralized, which can reduce fraud, make claim processing smoother, and help build trust in the insurance world.

The use of the Internet of Things (IoT) and telematics is particularly noticeable in health and auto insurance. Things like wearable devices and connected cars provide real-time data. This allows insurers to offer policies based on how you use your car or how healthy your lifestyle is, with rewards for safe driving or healthy habits. As insurers handle a lot of personal data, protecting this information is crucial. They are investing in strong cybersecurity measures to keep this data safe and meet data protection laws, which helps build trust with customers in a digital age.

The emergence of InsurTech startups has brought fresh innovations to the industry. These startups use the latest technologies to create new insurance products, streamline customer experiences, and make claims processing more efficient. This has pushed traditional insurance companies to speed up their own use of digital technology.

Despite the many opportunities technology brings, it also presents challenges. Insurers face issues like overcoming resistance to digital adoption, meeting regulatory requirements, and the need for ongoing innovation. Looking forward, technologies like augmented reality and quantum computing are expected to bring further changes to the industry, offering new opportunities and challenges.

Technology is key in shaping both the present and future of the insurance industry. The move towards digital technology has not only made operations more efficient and improved customer experiences but also led to the development of new and innovative insurance models. As the industry evolves, insurers need to find the right balance between embracing new technologies, staying in line with regulations, and focusing on their customers to succeed in this changing environment.

Challenges in Insurance Sector

The insurance industry is going through a big change because of technology, and it's facing quite a few challenges along the way (Table 3-3). First, there's the big task of moving from old paper-based methods to digital systems. This needs to be done in a way that's easy for customers to use and access online. Then, using artificial intelligence and predictive analytics is a challenge. These technologies help make underwriting faster and risk assessments more accurate, but they're complex and need to be used correctly.

Table 3-3. List of Challenges Faced in Insurance Industry

Challenge Categories	Description
Digital Transformation Adoption	Transitioning from traditional paper-based methods to digital systems, focusing on user-friendliness and online accessibility.
AI and Predictive Analytics Integration	Implementing AI for automating underwriting and using predictive analytics for risk assessments, requiring accuracy and effective application.
Blockchain Technology Implementation	Navigating the complexity of blockchain to ensure secure and efficient transactions, reduce fraud, and build trust.
IoT and Telematics Adoption	Integrating IoT and telematics in health and auto insurance for real-time data collection, managing large data volumes, and addressing privacy concerns.
Cybersecurity and Data Protection	Investing in advanced cybersecurity measures to protect vast amounts of personal data and comply with data protection regulations.
Responding to InsurTech Disruption	Adapting to competition and innovation from InsurTech startups, requiring collaboration and acceleration of digital transformation.
Regulatory Compliance	Keeping up with and adhering to various, changing regulations across different regions while maintaining operational efficiency.
Overcoming Resistance to Digital Adoption	Addressing internal and external resistance to new digital methods from employees and customers accustomed to traditional practices.
Need for Continuous Innovation	Staying ahead with ongoing innovation in the face of rapid technological advancements, preparing for emerging technologies like augmented reality and quantum computing.
Balancing Tech Advancement with Customer-Centricity	Maintaining a focus on personalized customer service and trust-building while embracing technological advancements.

Blockchain technology is another area where insurers are finding challenges. It's great for making transactions secure and reducing fraud, but it's not simple to implement. When it comes to health and auto insurance, using the Internet of Things and telematics is changing things a lot. These technologies collect real-time data which helps offer personalized insurance plans, but managing all this data and keeping it private is tough.

Cybersecurity is a huge concern too. Insurers collect a lot of personal information, so they need to keep it really safe and follow strict data protection laws. The rise of InsurTech startups is also shaking things up. These new companies are bringing in fresh ideas and technology, so traditional insurance companies have to keep up and sometimes work with these startups.

Dealing with all the different rules and regulations in insurance is another big challenge. Insurers have to make sure they're following all these rules while still running their business efficiently. Resistance to adopting new digital methods can be a problem, both from people working in insurance and from customers who are used to the old ways.

Staying innovative is important too, especially as new technologies keep coming up. Insurers have to be ready to use things like augmented reality and quantum computing in the future. Finally, even with all this technology, they need to keep focusing on their customers, making sure they provide personal and trustworthy service.

So, while the insurance industry is definitely getting more advanced with technology, it's not without its challenges. They've got to find the right balance between using new tech, following rules, and keeping customers happy.

Each of these challenges requires strategic attention and innovation. Successfully navigating them is essential for insurance companies to remain efficient, compliant, and competitive in an ever-evolving industry.

Case Study 3

Allianz SE stands as a prime example of a global insurance and financial services giant tackling complex business dilemmas in the industry. Operating in over 70 countries and headquartered in Munich, Germany, Allianz serves a vast global customer base.

Dilemma 1: Regulatory Compliance vs. Operational Flexibility

Allianz navigates the intricate balance between strict regulatory compliance and maintaining operational flexibility across diverse global markets. The company has crafted a robust compliance framework, but constantly adapting to different international and local regulations presents an ongoing challenge.

Dilemma 2: Data Utilization vs. Privacy Concerns
Central to Allianz's operations is the management and analysis of extensive data. While the company invests in sophisticated data systems for risk assessment and creating personalized products, balancing effective data use with maintaining customer privacy is a critical dilemma.

Dilemma 3: Advancing Cybersecurity vs. Technological Accessibility
Prioritizing cybersecurity while promoting technological accessibility forms another dilemma for Allianz. The company implements top-tier security measures like encryption and security audits but must also ensure these measures don't hinder customer access to digital services.

Dilemma 4: Embracing Digital Innovation vs. Preserving Traditional Values
As Allianz adapts to technological advancements, incorporating AI, machine learning, and blockchain, the challenge lies in embracing these innovations without losing sight of the traditional values and practices that have defined their long-standing reputation.

Dilemma 5: Streamlining Claims Processing vs. Ensuring Thoroughness
The company works to streamline claims processing for efficiency and customer satisfaction, yet it faces the dilemma of ensuring this efficiency doesn't compromise the thoroughness and accuracy of claims evaluation.

Dilemma 6: Enhancing Customer Experience vs. Cost Management
In the competitive insurance market, Allianz aims to enhance customer experience through personalized services and digital platforms. Balancing this with cost management, however, presents a significant challenge.

Dilemma 7: Combating Fraud vs. Customer Trust
Implementing rigorous fraud detection systems using AI and analytics is crucial for Allianz. The company must manage this while maintaining customer trust and ensuring legitimate claims are not unjustly impacted.

Dilemma 8: Risk Management vs. Competitive Risk Taking
Allianz employs advanced tools for risk assessment and management yet faces the dilemma of balancing cautious risk management with taking calculated risks to stay competitive in policy pricing and underwriting.

Dilemma 9: Navigating Market Volatility vs. Stable Growth
Managing exposure to market volatility and economic changes, while striving for stable growth and investment returns, is a constant balancing act for Allianz.

CHAPTER 3   CHALLENGES AND POTENTIAL APPLICATIONS OF GENERATIVE AI IN BFSI

Dilemma 10: Attracting Talent vs. Resource Allocation
The challenge of attracting and retaining top talent, especially in tech and data analytics, is juxtaposed against the need for prudent resource allocation and cost management.

Allianz SE's approach in dealing with these business dilemmas in the insurance sector showcases a strategic blend of innovation, adherence to traditional values, and a commitment to customer service. Successfully navigating these challenges underscores Allianz's role as a dynamic leader in the global insurance market, highlighting the need for agility and adaptability in a rapidly changing industry landscape.

Reimaging Landscape with Generative AI for Insurance Operations
Generative AI, with its advanced capabilities to simulate, predict, and automate, can offer transformative solutions to the operational challenges in the insurance sector. Here's how each challenge can be reimagined through the lens of Generative AI:

1. **Regulatory Compliance:** Generative AI can be employed to automatically interpret and adapt to changing regulations. It can generate compliance reports and documentation, ensuring that insurance companies remain up-to-date with regulatory changes while minimizing manual effort.

2. **Data Management and Analytics:** In data management, Generative AI can organize large datasets and generate insights for risk assessment, policy development, and customer segmentation. This not only improves efficiency but also enhances decision-making accuracy.

3. **Cybersecurity:** Generative AI can be pivotal in enhancing cybersecurity. It can simulate cyberattacks to identify potential vulnerabilities and generate robust defense mechanisms, thus proactively safeguarding sensitive data against emerging threats.

4. **Technological Adaptation:** For integrating new technologies, Generative AI can simulate various integration scenarios, helping to identify the most efficient paths for incorporating new systems with minimal disruption to existing operations.

5. **Claims Processing Efficiency:** Generative AI can automate and optimize the claims processing workflow. It can generate predictive models to streamline claim validation and assessment, reducing processing times and enhancing customer satisfaction.

6. **Customer Experience and Engagement:** Generative AI can personalize customer interactions by generating tailored communication and service recommendations. It can also create dynamic customer engagement strategies, improving the overall customer experience and loyalty.

7. **Fraud Detection and Prevention:** In fraud detection, Generative AI can simulate fraudulent activities to train and enhance fraud detection systems. Its ability to learn and adapt from new patterns can significantly improve the accuracy and efficiency of fraud prevention mechanisms.

8. **Risk Assessment and Management:** Generative AI can revolutionize risk assessment by generating predictive models based on vast datasets. These models can forecast potential risks and inform more accurate underwriting and pricing strategies.

9. **Market Volatility and Economic Changes:** To address market volatility, Generative AI can analyze market trends and economic data, generating predictive insights that help in formulating responsive strategies to mitigate financial impacts.

10. **Talent Acquisition and Retention:** In the realm of HR, Generative AI can automate the recruitment process, from generating job descriptions to initial candidate screening. It can also identify skill gaps within the organization and suggest tailored training and development programs.

3.6. Blueprint for Success: A Comprehensive Checklist for Enterprise Integration of Generative AI

As enterprises increasingly embrace digital transformation, the integration of Generative AI into business operations emerges as a pivotal strategic move. This checklist serves as a fundamental guide for companies preparing to navigate the complexities of implementing Generative AI. It outlines a structured approach, encompassing all critical facets from strategic alignment and feasibility assessment to compliance, risk management, and beyond. Designed to assist decision-makers and stakeholders, this comprehensive checklist ensures that each step towards integrating Generative AI is taken with informed caution and precision, aligning closely with the enterprise's overarching goals and the dynamic landscape of technological innovation. It is an indispensable tool for any enterprise embarking on the transformative journey of incorporating Generative AI into their operational fabric.

Implementing Generative AI in an enterprise setting is a complex process that requires meticulous planning and consideration of various factors. Below is a detailed checklist for enterprises considering the integration of Generative AI into their operations:

1. **Strategic Alignment**
    - Assess alignment with business goals and objectives.
    - Determine specific use cases for Generative AI in the enterprise.

2. **Feasibility Study**
    - Conduct a feasibility analysis for the proposed implementation.
    - Evaluate the existing infrastructure's readiness for Generative AI integration.

3. **Technology Assessment**
    - Identify the Generative AI technologies best suited for enterprise needs.
    - Evaluate the scalability and performance of chosen technologies.

4. **Data Management**
    - Ensure access to quality data sources necessary for training models.
    - Implement robust data governance policies and practices.

5. **Compliance and Ethics**
    - Review and adhere to relevant legal and regulatory compliance requirements.
    - Develop ethical guidelines for Generative AI use, addressing biases and privacy.

CHAPTER 3 CHALLENGES AND POTENTIAL APPLICATIONS OF GENERATIVE AI IN BFSI

6. **Risk Management**

    - Identify and assess potential risks associated with Generative AI deployment.

    - Develop risk mitigation and management strategies.

7. **Budget and Resource Allocation**

    - Allocate budget for technology acquisition, implementation, and maintenance.

    - Plan for human resources needed for development, operation, and oversight.

8. **Stakeholder Engagement**

    - Communicate with stakeholders to align expectations and garner support.

    - Provide necessary training and awareness programs for employees.

9. **Pilot Testing**

    - Conduct pilot tests to evaluate the effectiveness and impact of Generative AI solutions.

    - Collect feedback and make necessary adjustments before full-scale implementation.

10. **Integration and Deployment**

    - Develop a detailed plan for the integration of Generative AI into existing systems.

    - Execute deployment with continuous monitoring for performance and issues.

11. **Performance Evaluation and Monitoring**

    - Establish metrics for evaluating the performance and effectiveness of Generative AI solutions.

    - Implement continuous monitoring mechanisms to track performance and outcomes.

CHAPTER 3   CHALLENGES AND POTENTIAL APPLICATIONS OF GENERATIVE AI IN BFSI

12. **Continuous Improvement**

    - Gather insights and learnings from ongoing operations.
    - Update and improve Generative AI models and strategies regularly.

13. **Future Scalability**

    - Plan for future scalability and adaptability of Generative AI solutions.
    - Stay informed about emerging trends and advancements in Generative AI.

This checklist provides a comprehensive framework for enterprises to methodically approach the implementation of Generative AI. It ensures that critical aspects such as strategic alignment, technology assessment, compliance, and risk management are adequately addressed, paving the way for a successful and sustainable integration of Generative AI into business operations.

In each case, Generative AI not only offers solutions to the specific operational challenges but also adds value by enhancing efficiency, accuracy, and predictive capabilities. Its deployment in the insurance sector could lead to significant improvements in operational processes, customer service, and strategic decision-making.

The concluding part of the chapter emphasizes the pivotal role of Generative AI in reshaping the Banking, Financial Services, and Insurance sector. It highlights how this advanced technology is not only streamlining current operations but also paving the way for future innovations. The chapter concludes by stressing the importance of adapting to these technological advancements while maintaining a strong focus on customer-centric approaches, ensuring that the sector remains dynamic and responsive to evolving market needs and technological possibilities.

3.7. Summary

- The BFSI sector is undergoing a significant digital transformation, with a shift towards online and mobile banking platforms that enhance customer accessibility and convenience while presenting challenges in system integration and continuity.

CHAPTER 3 CHALLENGES AND POTENTIAL APPLICATIONS OF GENERATIVE AI IN BFSI

- The integration of AI and machine learning in banking is crucial for personalized customer services, risk assessment, fraud detection, and predictive analytics, requiring significant investment and expertise.

- As banking becomes more digital, cybersecurity has emerged as a critical focus, with banks needing to continually update their defenses to protect sensitive financial data from increasingly sophisticated cyber threats.

- The BFSI sector faces the challenge of adhering to a complex and ever-changing regulatory environment, necessitating robust compliance frameworks to ensure operational integrity across different regions.

- Effective risk management remains vital, with banks needing to develop flexible strategies to manage credit risk, market volatility, and operational issues while maintaining service quality and financial performance.

- Incorporating new technologies like AI, blockchain, and cloud computing into existing banking operations is complex and costly, requiring careful planning to avoid disruption and ensure seamless integration.

- Enhancing customer experience through personalization and efficient service delivery is key to retaining and attracting customers in a competitive market, necessitating ongoing innovation and technology adoption.

- The management and utilization of vast amounts of data are crucial for informed decision-making in the BFSI sector, with challenges in ensuring data privacy, security, and effective analytics.

- The sophistication of financial fraud necessitates the continuous enhancement of detection and prevention mechanisms, using advanced technologies to protect customer transactions and bank operations.

- The rapid technological changes in the BFSI sector have made it increasingly challenging to attract and retain skilled professionals, particularly in areas like technology and data analysis, requiring competitive strategies to build and maintain a capable workforce.

PART II

Generative AI's Transformative Impact on Financial Verticals

CHAPTER 4

Transforming Banking: The Next Frontier

4.1. Introduction

In an era marked by rapid technological advancements, the Banking, Financial Services, and Insurance (BFSI) sector is undergoing profound transformations driven by generative AI. This chapter delves into the foundational insights of how generative AI is reshaping pivotal areas within the BFSI landscape. From enhancing transactional operations to revolutionizing customer engagement, credit assessment, and security protocols, generative AI is at the forefront of this evolution. Additionally, this chapter explores AI-driven innovations in investment strategies and regulatory compliance, providing a comprehensive view of the current and future impacts of AI technologies on the BFSI sector.

Generative AI, a subset of artificial intelligence, leverages machine learning models to generate new content, ideas, and solutions based on existing data. This technology goes beyond traditional AI applications by offering creative problem-solving capabilities, making it a powerful tool for the BFSI industry. Banks and financial institutions are increasingly adopting generative AI to streamline operations, enhance customer experiences, and ensure regulatory compliance. For instance, AI models can analyze vast datasets to provide personalized financial advice, detect fraudulent activities in real time, and automate complex reporting tasks.

The integration of generative AI in BFSI is not merely an enhancement but a fundamental shift in how financial services are delivered and managed. This transformation is evident in several key areas. Transactional operations are becoming more efficient through AI-driven automation, reducing the need for manual intervention and minimizing errors. Customer support and engagement are being redefined with AI-powered chatbots and virtual assistants that offer personalized interactions and solutions. Credit assessment processes are evolving, with AI providing more accurate and inclusive evaluations of creditworthiness.

Moreover, generative AI is enhancing fraud detection and security protocols by identifying subtle patterns and anomalies that traditional methods might miss. Investment strategies are also benefiting from AI-driven insights, enabling more informed decision-making and risk management. In the realm of regulatory compliance, AI is streamlining reporting processes and ensuring adherence to complex regulatory requirements.

As we explore these transformative applications, this chapter provides a systematic analysis of how generative AI is revolutionizing the BFSI sector. Through real-world examples, industry studies, and expert insights, we will uncover the profound impacts of AI-driven innovations on transactional operations, customer engagement, credit processes, security, investment strategies, and regulatory compliance. This discourse aims to provide a comprehensive understanding of the current landscape and envision the future prospects of the BFSI sector powered by generative AI.

Navigating Modernization in the Banking Industry

The banking industry faces an urgent need to modernize, driven by the demand for enhanced customer experiences, operational agility, competition from FinTech and digital banks, and evolving regulatory requirements. Legacy systems, such as on-premises servers and interdepartmental hand-offs, excel in handling high volumes and mission-critical functions but fall short in meeting contemporary standards for banking operations.

Customer Experience (CX) Challenges

Traditional core banking systems (CBS) often fail to deliver the seamless, fast, and personalized services that digital-savvy customers expect. Modern customers interact with banks through multiple touchpoints and channels, requiring secure and context-aware services. A superior employee experience is crucial for delivering optimal customer experiences. Achieving "customer delight" in modern banking necessitates radical changes, including digital adoption, workforce upskilling/reskilling, harnessing data insights, and building a customer-centric ecosystem.

Operational Agility

Operational agility is essential for modern banking, encompassing tech stack modernization, policy updates, process improvements, and skillset enhancements. Banks must be responsive and adaptable to market demands, delivering just-in-time services with high quality while staying compliant with changing regulations. Achieving operational dexterity involves shifting organizational culture and capabilities alongside digital solutions.

Challenges to Modernization

Despite recognizing the need for modernization, banks face significant challenges:

1. **Legacy Systems**: Core banking systems are monolithic, with entrenched technology stacks, policies, and processes developed over decades. This setup relies on predictable operations like batching and cyclical reviews, which can overlook the need for adaptive, agile approaches. Regulatory mandates requiring immediate action might be delayed until end-of-day batch processing, and the lack of data insights can hinder process improvements, increasing enterprise risks and delivery delays.

2. **Siloed Sub-systems**: Traditional banking operations often feature siloed sub-systems that do not communicate in real time. This lack of synergy and data insights can lead to inefficiencies and poor customer experiences. For example, customer data captured through a web form may not be reconciled in the database, leading to duplicate efforts and a broken experience. Additionally, a customer's preference collected at one touchpoint might not be met at another, resulting in dissatisfaction.

3. **Reactive Approach**: Traditional banking typically adopts a reactive approach, initiating actions or decisions only after events occur. In dynamic market and regulatory environments, this retrospective mode offers limited real-time agility.

4. **Technological Upgrades**: Upgrading or replacing technology in banking requires a domain-centric consultative approach. According to McKinsey, core banking system modernization can be achieved through complete replacement, implementing greenfield core banking suites, or revamping and decomposing existing systems. Each approach has pros and cons; full replacement and greenfield options are costly and risky due to large-scale migrations. Revamping requires careful assessment of components and inner workings to ensure effective execution

and change management. For example, integrating legacy systems with niche service packages or utilities like data analytics programs or automation systems can be challenging, even with APIs. A combination of domain-led approach and implementation skills is crucial for successful modernization.

4.2. Imperative Actions for Banking Modernization with Generative AI

Generative AI is transforming the banking industry (Figure 4-1) offering innovative solutions to enhance customer experience, operational agility, and regulatory compliance. Here are key interventions that can help banks leverage this technology to modernize their operations:

1. **Harnessing Customer and Operational Insights with Generative AI**

    Generative AI can process and analyze the massive amounts of data generated by the banking sector to provide valuable insights. By 2020, data generation in banking saw a nearly 700% increase per second, encompassing customer preferences, transaction details, revenue data, KYC information, production data, and risks.

    Generative AI enhances data analytics by transforming raw data into actionable insights via SMART dashboards. The process involves several steps:

    - **Data Collection**: Generative AI gathers data from various sources efficiently.
    - **Data Collation**: It combines data into comprehensive datasets.
    - **Data Cleaning and Sorting**: AI ensures data quality by removing errors and inconsistencies.
    - **Data Entry and Indexing**: AI automates data input and organization for efficient retrieval.

- **Database Rules Processing**: It manages data according to predefined rules.

- **User Interface Mapping**: AI connects data to user interfaces seamlessly.

- **Data Analysis and Dissemination**: It interprets data and shares insights.

- **Data Visualization**: AI-powered SMART dashboards display data insights visually for better comprehension.

Modern data analytics systems use APIs to ingest and store data in cloud repositories, applying AI/ML algorithms for real-time insights. Tools like optical character recognition (OCR) and robotic process automation (RPA) accelerate data collection and collation from various formats, including digital documents, print copies, and handwritten notes.

2. **Implementing Digital Solutions with Generative AI**

   Generative AI can significantly enhance traditional banking systems by automating processes and providing real-time insights. These AI-powered solutions can monitor financial transactions for compliance with anti-money laundering (AML) regulations, analyze and visualize data, and transform traditional BSA/AML/CFT compliance capabilities through real-time transaction monitoring, watchlist screening, and KYC risk profiling.

   The implementation of generative AI can focus on revamping and decomposing legacy systems to minimize risks associated with large-scale replacements. Bolt-on solutions and API-based approaches provide flexibility and customization for traditional banking systems that offer limited integration options.

3. **Workforce Skilling and Knowledge Management with Generative AI**

   Generative AI can play a crucial role in upskilling and reskilling the workforce, aligning human skills with technological advancements. Effective knowledge management, encompassing process workflows, operational expertise, and risk oversight, is essential for supporting banking modernization.

AI-powered digital learning platforms with curated knowledge repositories and domain-led digital libraries can deliver focused training programs. Customizable digital courseware can support knowledge management by creating and disseminating digital business process blueprints maintained in a centralized, up-to-date content library.

4. **Flexible Operating Models with Generative AI**

   Generative AI can enhance flexible banking operations through a modular and function-oriented design. This approach allows specific delivery parameters to be configured across various banking verticals, such as service support, treasury operations, commercial lending, and retail servicing.

   For instance, AI can streamline KYC and Customer Due Diligence processes across different banking functions, improving flexibility and agility. Generative AI can also support predictive analytics, enabling banks to anticipate market trends and customer behaviors, thus making proactive decisions and mitigating risks.

Reimagining Banking Transactions with Generative AI

Generative AI has the potential to revolutionize banking transactions by offering hyper-personalized services, enhancing predictive analytics, and automating customer support.

- **Hyper-personalized Services**: AI analyzes vast amounts of data to tailor banking services to individual customer needs, enhancing satisfaction and loyalty.

- **Predictive Analytics**: AI predicts market trends and customer behaviors, enabling banks to make proactive decisions and mitigate risks.

- **Automated Customer Support**: AI-powered chatbots and virtual assistants provide real-time support, resolving customer queries efficiently and enhancing the overall customer experience.

1. **Redesigning Front Office Operations: The Role of Call Centers, Online Queries, and Robotic Automation**

   The transformation of front office operations in the banking industry is increasingly driven by the integration of Large Language Models (LLMs) into call centers, online query systems, and robotic process automation (RPA). These technologies not only enhance customer experience but also improve operational efficiency by automating and streamlining interactions.

   LLMs in Call Centers and Online Queries: Revolutionizing Customer Interaction

   LLMs are playing a pivotal role in modernizing front office operations by enabling more intelligent and responsive customer service systems. According to research by McKinsey & Company, integrating LLMs into call centers can significantly reduce customer wait times and enhance service accuracy. LLM-powered systems can understand and process natural language queries, allowing for more precise routing of customer inquiries to the most suitable agents. This leads to quicker and more effective problem resolution.

   For example, HSBC's implementation of an AI-driven call center solution, which utilizes LLMs for natural language processing (NLP), has resulted in a 40% reduction in call handling times and a 30% improvement in customer satisfaction scores (McKinsey & Company, 2022). LLMs, in this context, analyze customer queries in real time, providing relevant responses and even handling complex inquiries autonomously. Additionally, online query systems powered by LLMs enable customers to resolve issues through self-service portals, reducing the load on call centers and enhancing overall service delivery.

   By leveraging LLMs, banks are not only optimizing their front office operations but are also setting new standards for customer service, driving both cost savings and customer satisfaction.

CHAPTER 4   TRANSFORMING BANKING: THE NEXT FRONTIER

Robotic Automation in Front Office Operations

Robotic Process Automation (RPA) and AI-driven robots are increasingly being deployed in front office operations. These technologies automate routine tasks, allowing human agents to focus on more complex customer interactions. According to Deloitte's 2023 report on banking automation, RPA can handle tasks such as data entry, transaction processing, and compliance checks, reducing error rates and operational costs.

A real-time example of successful robotic automation is the implementation of "Pepper" robots by Mitsubishi UFJ Financial Group (MUFG). These humanoid robots assist customers in branches by answering basic queries, guiding them through transactions, and even providing information on financial products. This initiative not only enhances customer experience but also allows bank staff to concentrate on advisory roles (Deloitte, 2023).

Benefits and Challenges

The primary benefits of redesigning front office operations include improved efficiency, cost reduction, and enhanced customer satisfaction. A study by Forrester Research indicates that banks utilizing AI and RPA in their front offices report a 25–30% reduction in operational costs and a significant increase in customer retention rates.

However, the integration of these technologies also presents challenges. Data privacy and security are major concerns, as highlighted by a report from the International Data Corporation (IDC). Ensuring that AI systems comply with regulatory standards and protect sensitive customer data is paramount. Moreover, there is a need for continuous monitoring and updating of these systems to adapt to evolving customer needs and regulatory requirements.

Future Prospects

The future of front office operations in banking lies in the seamless integration of AI, RPA, and human agents. Research by Gartner suggests that by 2025, 75% of customer interactions in banks will be handled by AI-powered systems, with human agents stepping in only for complex issues. This hybrid model promises to deliver superior customer experiences while maintaining high operational efficiency.

Redesign of front office operations through call centers, online queries, and robotic automation offers significant advantages for banks. By leveraging these technologies, banks can improve service delivery, reduce costs, and enhance customer satisfaction. However, it is crucial to address the challenges of data privacy and regulatory compliance to fully realize these benefits.

2. **Integrating Generative AI for Product Descriptions, Online KYC, Transaction Handling, and Product Pitching Through Chatbots**

The integration of generative AI in the banking sector is transforming various aspects of front office operations, including product descriptions, online Know Your Customer (KYC) processes, transaction handling, and product pitching through chatbots. These advancements are not only enhancing operational efficiency but also improving customer experience and regulatory compliance.

Product Descriptions with Generative AI

Generative AI can create detailed and personalized product descriptions tailored to individual customer preferences. This technology uses natural language processing (NLP) to analyze customer data and generate customized descriptions that resonate with specific needs. For example, JPMorgan Chase has implemented AI-driven systems that personalize product offerings based on customer profiles and transaction history, enhancing engagement and conversion rates (Forbes, 2023).

According to a study by Accenture, banks utilizing AI for content creation experience a 20% increase in customer engagement compared to traditional methods. Automating the generation of product descriptions ensures consistency, accuracy, and appeal across all customer touchpoints.

Online KYC Processes

The integration of generative AI in online KYC processes offers significant improvements in efficiency and accuracy. Traditional KYC methods are often time-consuming and prone to human error. Generative AI streamlines this process by automating the verification of customer identities, reducing the time required to onboard new clients.

HSBC's implementation of an AI-powered KYC solution demonstrates this efficiency. The system uses AI to analyze and verify customer documents in real time, significantly reducing the onboarding time from days to minutes (Cointelegraph, 2023). Moreover, the use of AI ensures higher accuracy in document verification, minimizing the risk of fraud and enhancing compliance with regulatory standards (NS Banking, 2023).

Handling Transactions

Generative AI also plays a critical role in handling transactions. By automating routine tasks such as data entry and transaction monitoring, AI reduces the operational burden on bank staff. AI algorithms can detect anomalies in real time, providing enhanced security and fraud detection capabilities.

For instance, Bank of America's Erica, an AI-driven virtual assistant, handles over 50 million customer interactions annually, including transaction-related queries and tasks (Bank of America, 2023). Erica uses machine learning to continuously improve its responses, offering customers a seamless and efficient way to manage their banking transactions.

Product Pitching Through Chatbots

Generative AI-powered chatbots are revolutionizing how banks pitch products to customers. These chatbots use NLP to understand customer inquiries and provide personalized product recommendations in real time. By analyzing customer data, chatbots can identify potential needs and suggest relevant financial products.

A notable example is the chatbot deployed by Wells Fargo. This AI-driven assistant engages customers in meaningful conversations, understands their financial needs, and pitches suitable products such as loans, credit cards, and investment options (Gartner, 2023). The chatbot's ability to provide instant, personalized recommendations improves customer satisfaction and drives product adoption.

Real-Time Examples and Industry Impact

The practical application of generative AI in banking is best illustrated by its impact on customer experience and operational efficiency. According to a report by Deloitte, banks using AI-driven solutions see a 30% reduction in customer service costs and a 40% increase in customer satisfaction (Deloitte, 2023). These metrics highlight the tangible benefits of integrating AI technologies in front office operations.

Furthermore, the use of AI in banking supports regulatory compliance by ensuring accurate and timely processing of customer data. AI systems can continuously monitor transactions and flag suspicious activities, enhancing the bank's ability to comply with anti-money laundering (AML) regulations and other compliance requirements (NS Banking, 2023).

Integrating generative AI into product descriptions, online KYC processes, transaction handling, and product pitching through chatbots represents a significant advancement in the banking industry. These technologies not only enhance operational efficiency but also provide a superior customer experience.

CHAPTER 4   TRANSFORMING BANKING: THE NEXT FRONTIER

By leveraging AI, banks can stay competitive in a rapidly evolving financial landscape, ensuring they meet the expectations of modern customers while maintaining strict regulatory compliance.

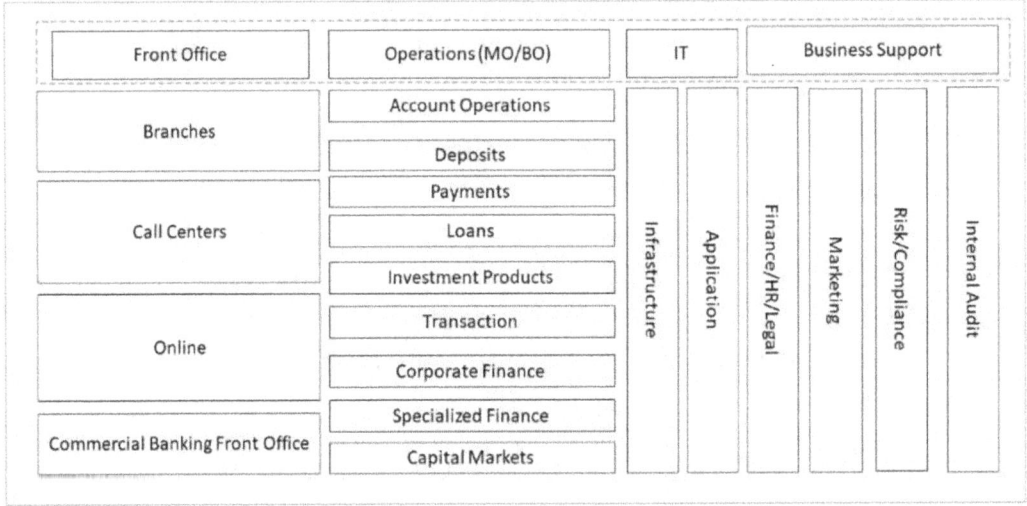

Figure 4-1. Banking Operations

3. **Restructuring IT Infrastructure and Applications**

   Generative AI is revolutionizing the restructuring of IT infrastructure and applications across various industries. By leveraging advanced machine learning algorithms, generative AI enhances efficiency, reduces costs, and optimizes resource utilization in IT operations. This article explores how generative AI is applied in IT restructuring, highlighting real-time examples and its impact on the industry.

   Enhancing IT Infrastructure with Generative AI

   Generative AI plays a crucial role in automating and optimizing IT infrastructure management. It can design, deploy, and manage complex IT systems by predicting future states and generating optimal configurations. This capability is particularly beneficial in cloud computing, where resource allocation and management are critical.

For instance, Google Cloud uses generative AI to optimize its data centers. The AI models predict server failures, optimize cooling systems, and manage power consumption, leading to significant cost savings and improved efficiency (Google Cloud, 2023). According to a study by Gartner, companies implementing AI-driven IT infrastructure management see a 30% reduction in operational costs and a 40% increase in system reliability (Gartner, 2023).

Application Modernization Through Generative AI

Application modernization is another area where generative AI is making a significant impact. Legacy applications often require updates to improve performance, security, and compatibility with modern IT environments. Generative AI can analyze existing codebases, identify areas for improvement, and generate new code that enhances functionality and efficiency.

A notable example is IBM's use of AI to modernize legacy applications for its clients. IBM's generative AI tools analyze millions of lines of code, identify optimization opportunities, and generate new, more efficient code. This process not only speeds up modernization efforts but also reduces the risk of errors and security vulnerabilities (IBM, 2022).

Real-Time Example: Microsoft's Azure AI

Microsoft's Azure AI provides a comprehensive suite of AI services that leverage generative AI to enhance IT infrastructure and applications. Azure AI can automatically scale resources based on demand, optimize virtual machine deployments, and manage network configurations. This automation leads to improved resource utilization and reduced operational overhead.

One real-time example is Microsoft's collaboration with Walmart. Walmart uses Azure AI to manage its extensive IT infrastructure, optimizing everything from server allocations to network configurations. This partnership has resulted in a 25% reduction in infrastructure costs and a significant improvement in system performance (Microsoft, 2023).

Generative AI in IT Security

Generative AI also enhances IT security by identifying potential threats and generating proactive solutions. AI models can analyze network traffic, detect anomalies, and predict future attacks. This proactive approach to security helps organizations protect their IT infrastructure and sensitive data.

According to a report by McKinsey & Company, organizations using AI-driven security solutions experience a 50% reduction in successful cyber-attacks and a 60% improvement in threat detection and response times (McKinsey & Company, 2023). These improvements are critical in today's digital landscape, where cyber threats are becoming increasingly sophisticated.

Challenges and Future Prospects

While generative AI offers numerous benefits, its implementation is not without challenges. Organizations must ensure they have the necessary data infrastructure and expertise to leverage AI effectively. Additionally, ethical considerations around data privacy and AI decision-making must be addressed.

Despite these challenges, the future of generative AI in IT restructuring is promising. As AI technologies continue to evolve, their capabilities in automating and optimizing IT operations will only improve. According to a report by Forrester Research, the global market for AI in IT operations is expected to grow at a compound annual growth rate (CAGR) of 25% over the next five years (Forrester Research, 2023).

Generative AI is transforming the restructuring of IT infrastructure and applications, offering significant benefits in terms of efficiency, cost reduction, and security. Real-time examples from industry leaders like Google, IBM, and Microsoft demonstrate the practical applications and impact of this technology. As organizations continue to adopt and integrate generative AI, the future of IT operations looks increasingly automated and optimized, promising even greater advancements in the years to come.

By understanding and leveraging generative AI, organizations can stay ahead in the competitive IT landscape, ensuring they meet the demands of modern business environments while maintaining robust and secure IT systems.

4.3. Designing Banking Applications: Leveraging Generative AI for Innovation

The banking industry is rapidly evolving, driven by technological advancements and changing customer expectations. Designing banking applications that are secure, efficient, and user-friendly is paramount. Generative AI plays a pivotal role in this transformation, enabling banks to develop sophisticated applications that enhance customer experience, streamline operations, and ensure regulatory compliance. This article explores how generative AI is utilized in designing banking applications, supported by real-time examples and industry insights.

Enhancing User Experience

Generative AI significantly improves the user experience (UX) in banking applications by personalizing services and interfaces. AI algorithms analyze user behavior and preferences to generate customized interfaces that enhance usability and engagement.

For instance, Bank of America's AI-driven virtual assistant, Erica, provides personalized financial advice, transaction notifications, and budgeting tips based on user data (Bank of America, 2023). This level of personalization helps in creating a more engaging and satisfying user experience, leading to higher customer retention rates.

Streamlining Operations

Generative AI optimizes banking operations by automating routine tasks and improving process efficiency. AI-driven applications can handle various tasks, such as transaction processing, fraud detection, and customer support, with minimal human intervention.

A notable example is the use of AI by JPMorgan Chase for its Contract Intelligence (COiN) platform, which automates the review of legal documents and extracts critical data points in seconds, a task that previously took thousands of hours annually (JPMorgan Chase, 2023). This automation not only reduces operational costs but also minimizes errors and enhances regulatory compliance.

Security and Fraud Detection

Security is a critical concern in banking applications. Generative AI enhances security by continuously monitoring transactions and identifying suspicious activities in real time. AI algorithms can detect patterns and anomalies that may indicate fraudulent behavior, enabling banks to take proactive measures.

HSBC's implementation of an AI-powered anti-money laundering (AML) system exemplifies this application. The system uses machine learning to monitor transactions and identify high-risk profiles, significantly reducing false positives and improving detection accuracy (Cointelegraph, 2023).

Personalized Financial Services

Generative AI enables banks to offer personalized financial services tailored to individual customer needs. By analyzing vast amounts of customer data, AI can generate insights that help in creating customized investment plans, loan offers, and insurance products.

For example, Wells Fargo uses AI-driven chatbots to engage with customers, understand their financial needs, and recommend suitable products (Gartner, 2023). This personalized approach enhances customer satisfaction and drives product adoption.

Development of Intelligent Chatbots

Intelligent chatbots powered by generative AI are transforming customer support in banking applications. These chatbots can handle a wide range of customer queries, provide instant responses, and offer personalized financial advice, thus improving customer service efficiency.

The chatbot deployed by TD Bank is a prime example. It uses AI to provide real-time assistance to customers, helping them with tasks such as account inquiries, transaction history, and financial planning (Forrester Research, 2023). This not only enhances the customer experience but also reduces the workload on human agents.

Case Study: Microsoft's Azure AI in Banking

Microsoft's Azure AI platform offers a comprehensive suite of AI services that are extensively used in banking applications. Azure AI helps banks in automating operations, enhancing security, and delivering personalized customer experiences.

In a collaboration with Standard Chartered, Azure AI is used to streamline the bank's operations, improve risk management, and deliver personalized services to clients (Microsoft, 2023). This partnership highlights the potential of AI in transforming banking operations and improving service delivery.

Challenges and Considerations

While generative AI offers numerous benefits, its implementation in banking applications comes with challenges. Ensuring data privacy, managing AI biases, and maintaining regulatory compliance are critical considerations. Banks must invest in robust data governance frameworks and continuously monitor AI systems to mitigate these risks.

Generative AI is revolutionizing the design and development of banking applications. By enhancing user experience, streamlining operations, improving security, and offering personalized services, AI-driven applications are setting new standards in the banking industry. Real-time examples from industry leaders like Bank of America, JPMorgan Chase, and HSBC demonstrate the transformative potential of generative AI. As banks continue to adopt and integrate AI technologies, the future of banking applications looks increasingly innovative, efficient, and customer-centric.

By leveraging generative AI, banks can not only meet the evolving needs of their customers but also stay ahead in the competitive financial landscape, ensuring robust and secure banking systems.

4.4. Redefining Bank Business Support Functions with Generative AI

Generative AI is transforming business support functions within banks, including finance, human resources (HR), and legal departments. By leveraging advanced machine learning algorithms, generative AI enhances efficiency, accuracy, and decision-making processes, leading to significant improvements in these critical areas.

Finance

In the finance department, generative AI can streamline numerous processes, from financial planning and analysis to risk management and reporting. AI algorithms can analyze vast datasets to generate financial forecasts, identify trends, and detect anomalies. This capability enhances the accuracy of financial projections and supports better decision-making.

For example, AI-driven financial planning tools can automate the preparation of budgets and financial reports. According to a study by PwC, companies that adopt AI for financial planning and analysis report a 20–30% reduction in time spent on these tasks and a 15–20% improvement in forecast accuracy (PwC, 2023). Additionally, AI can assist in risk management by identifying potential risks and suggesting mitigation strategies. A notable example is BlackRock's Aladdin platform, which uses AI to provide comprehensive risk analytics and portfolio management services (BlackRock, 2023).

Human Resources

Generative AI has the potential to revolutionize HR functions by automating recruitment processes, enhancing employee engagement, and improving talent management. AI-powered recruitment tools can screen resumes, conduct initial interviews, and assess candidate fit based on predefined criteria. This automation speeds up the hiring process and reduces biases.

A real-world example is Unilever's use of AI in its recruitment process. The company employs AI-driven tools to analyze video interviews and predict candidate success, resulting in a 50% reduction in time-to-hire and a 16% increase in diversity (Harvard Business Review, 2022). Furthermore, AI can support employee engagement by providing personalized career development plans and training recommendations, enhancing employee satisfaction and retention.

Legal

In the legal domain, generative AI can significantly enhance contract management, compliance monitoring, and legal research. AI-powered tools can analyze legal documents, identify key clauses, and suggest modifications, reducing the time and effort required for contract review and drafting.

For instance, JPMorgan Chase's COiN (Contract Intelligence) platform automates the review of legal documents, extracting important data points in seconds—a task that previously took thousands of hours annually (JPMorgan Chase, 2023). Additionally, AI can aid in compliance by continuously monitoring regulatory changes and ensuring that the organization's policies and procedures are up-to-date. This proactive approach helps banks avoid regulatory penalties and maintain compliance with evolving laws.

Benefits and Challenges

The integration of generative AI in business support functions offers numerous benefits, including increased efficiency, cost savings, and enhanced decision-making. According to a report by Deloitte, organizations using AI in their support functions experience a 30% reduction in operational costs and a significant improvement in service delivery (Deloitte, 2023).

However, the implementation of AI also presents challenges. Ensuring data privacy and security, managing AI biases, and maintaining regulatory compliance are critical considerations. Organizations must invest in robust data governance frameworks and continuously monitor AI systems to mitigate these risks.

The future of generative AI in business support functions is promising. As AI technologies continue to evolve, their capabilities in automating and optimizing processes will further enhance the efficiency and effectiveness of finance, HR, and legal departments. According to a study by Forrester Research, the global market for AI in business support functions is expected to grow at a compound annual growth rate (CAGR) of 25% over the next five years (Forrester Research, 2023).

Generative AI is redefining the functionality of business support functions in banks. By enhancing efficiency, accuracy, and decision-making in finance, HR, and legal departments, AI-driven solutions are setting new standards in the industry. Real-time examples from leading organizations such as BlackRock, Unilever, and JPMorgan Chase demonstrate the transformative potential of generative AI. As banks continue to adopt and integrate AI technologies, they can achieve significant operational improvements and stay competitive in the rapidly evolving financial landscape.

Redefining Bank Marketing Functions with Generative AI

Generative AI is revolutionizing marketing functions within the banking industry, offering new avenues for personalized customer engagement, data-driven decision-making, and campaign optimization. This article explores how generative AI is being utilized in bank marketing functions, with real-world examples and insights from industry studies.

Personalized Customer Engagement

Generative AI enhances personalized customer engagement by analyzing vast amounts of customer data to generate tailored marketing content. This technology allows banks to create personalized email campaigns, social media posts, and advertisements that resonate with individual customer preferences and behaviors.

For example, the Royal Bank of Canada (RBC) uses AI-driven marketing platforms to analyze customer data and deliver personalized messages. RBC's AI system tailors product recommendations based on customer transaction history, preferences, and financial goals, leading to a significant increase in customer engagement and conversion rates (RBC, 2023).

Data-Driven Decision-Making

Generative AI supports data-driven decision-making by providing deep insights into customer behavior, market trends, and campaign performance. AI algorithms can analyze customer interactions across multiple channels, identify patterns, and predict future behaviors, enabling marketers to make informed decisions.

A notable example is the use of AI by JP Morgan Chase in its marketing strategies. The bank employs AI to analyze data from various sources, such as social media, transaction history, and customer feedback. This analysis helps JP Morgan Chase to understand customer needs better, optimize marketing campaigns, and allocate resources more effectively (JP Morgan Chase, 2023).

Campaign Optimization
Generative AI plays a crucial role in optimizing marketing campaigns. AI-driven platforms can test different campaign variables, such as messaging, timing, and channel, to determine the most effective combinations. This capability allows banks to run more efficient campaigns with higher return on investment (ROI).

For instance, ING Group uses AI to optimize its digital marketing campaigns. The bank's AI system tests various ad creatives and targeting strategies to identify the best-performing combinations. This approach has resulted in a 30% increase in campaign effectiveness and a 20% reduction in marketing costs (ING Group, 2023).

Real-Time Example: Wells Fargo's AI-Driven Marketing
Wells Fargo employs AI to enhance its marketing efforts, particularly in personalizing customer experiences and optimizing campaigns. The bank's AI system analyzes customer data to generate personalized marketing messages and offers. Wells Fargo also uses AI to monitor campaign performance in real time and adjust strategies accordingly.

This AI-driven approach has enabled Wells Fargo to achieve a higher level of customer satisfaction and engagement. According to a report by Forrester Research, Wells Fargo's use of AI in marketing has led to a 25% increase in customer acquisition and a 15% improvement in customer retention rates (Forrester Research, 2023).

Predictive Analytics
Predictive analytics powered by generative AI allows banks to anticipate customer needs and market trends. By analyzing historical data, AI can forecast future behaviors and preferences, enabling marketers to develop proactive strategies.

A real-world application of predictive analytics is seen in Citibank's marketing operations. Citibank utilizes AI to predict customer needs and tailor marketing efforts accordingly. This predictive capability helps the bank to offer relevant products and services at the right time, enhancing customer experience and increasing sales (Citibank, 2023).

While the benefits of generative AI in marketing are substantial, there are challenges to consider. Ensuring data privacy and security, managing AI biases, and maintaining regulatory compliance are critical issues that banks must address. Robust data governance frameworks and continuous monitoring of AI systems are essential to mitigate these risks.

Generative AI is redefining marketing functions in the banking sector by enabling personalized customer engagement, data-driven decision-making, and campaign optimization. Real-world examples from RBC, JP Morgan Chase, ING Group, Wells Fargo, and Citibank illustrate the transformative potential of AI in marketing. As banks continue to integrate AI technologies, they can expect significant improvements in marketing efficiency and effectiveness, ensuring they meet the evolving needs of their customers and stay competitive in the financial landscape.

Bank Risk and Compliance Functions with Generative AI

Generative AI is transforming risk and compliance functions in the banking industry, offering new ways to enhance efficiency, accuracy, and regulatory adherence. By leveraging advanced machine learning algorithms, generative AI can analyze vast datasets, predict risks, and ensure compliance with complex regulations. This article explores how generative AI is used in redefining risk and compliance in banks, supported by real-world examples and industry insights.

Enhancing Risk Management

Generative AI improves risk management by analyzing data from various sources to identify and predict potential risks. AI models can process vast amounts of data to detect patterns and anomalies that may indicate financial crimes, market risks, or operational risks.

For example, HSBC has implemented an AI-powered risk management system that uses machine learning to monitor transactions and identify high-risk activities. This system analyzes transaction data in real time, flagging suspicious activities and reducing false positives (HSBC, 2023). According to a report by McKinsey & Company, banks utilizing AI for risk management can reduce false positives by up to 60%, improving the accuracy and efficiency of their risk detection processes (McKinsey & Company, 2023).

Compliance Monitoring

Compliance monitoring is a critical function in banking, ensuring that institutions adhere to regulatory requirements. Generative AI can automate compliance checks, monitor regulatory changes, and ensure that banks remain compliant with evolving laws.

JP Morgan Chase's implementation of AI in compliance monitoring is a notable example. The bank uses AI to analyze regulatory documents and ensure that its operations comply with current regulations. This AI-driven approach has significantly reduced the time and cost associated with manual compliance checks (JP Morgan Chase, 2023). Furthermore, AI systems can continuously update compliance protocols as new regulations emerge, ensuring ongoing compliance.

Fraud Detection

Generative AI enhances fraud detection capabilities by analyzing transaction data and identifying fraudulent activities. AI models can detect subtle patterns that may be missed by traditional rule-based systems, providing a more robust fraud detection mechanism.

Wells Fargo's use of AI in fraud detection illustrates the effectiveness of this technology. The bank's AI system monitors customer transactions in real time, identifying and flagging potentially fraudulent activities. This proactive approach has resulted in a significant reduction in fraud losses and improved customer trust (Wells Fargo, 2023). According to Deloitte, banks using AI for fraud detection report a 50% reduction in fraud-related losses (Deloitte, 2023).

Predictive Analytics

Predictive analytics powered by generative AI allows banks to anticipate risks and take preventive measures. By analyzing historical data, AI models can forecast future risks and suggest mitigation strategies, enabling banks to stay ahead of potential threats.

Citibank's implementation of predictive analytics in its risk management processes is a prime example. The bank uses AI to predict market trends and identify potential risks, allowing it to make informed decisions and mitigate risks proactively (Citibank, 2023). This approach not only enhances risk management but also improves overall operational efficiency.

Real-Time Example: Standard Chartered's AI-Driven Compliance

Standard Chartered uses AI to enhance its compliance functions, particularly in anti-money laundering (AML) and Know Your Customer (KYC) processes. The bank's AI system analyzes customer data to identify suspicious activities and ensure compliance with AML regulations. This AI-driven approach has significantly improved the bank's ability to detect and prevent financial crimes (Standard Chartered, 2023).

According to a report by PwC, banks that adopt AI for compliance monitoring can reduce compliance costs by up to 30% and improve detection accuracy by 20% (PwC, 2023). These improvements are crucial in today's regulatory environment, where compliance requirements are continually evolving.

While generative AI offers substantial benefits, its implementation in risk and compliance functions also presents challenges. Ensuring data privacy and security, managing AI biases, and maintaining regulatory compliance are critical issues that banks must address. Robust data governance frameworks and continuous monitoring of AI systems are essential to mitigate these risks.

Generative AI is redefining risk and compliance functions in the banking industry by enhancing efficiency, accuracy, and regulatory adherence. Real-world examples from HSBC, JP Morgan Chase, Wells Fargo, Citibank, and Standard Chartered demonstrate the transformative potential of AI in these critical areas. As banks continue to adopt and integrate AI technologies, they can expect significant improvements in risk management and compliance, ensuring they meet the evolving regulatory requirements and stay competitive in the financial landscape.

Internal Audit Functions with Generative AI

Generative AI is revolutionizing internal audit functions in the banking sector, providing enhanced capabilities for risk assessment, compliance monitoring, and operational efficiency. By leveraging advanced machine learning algorithms, generative AI can analyze vast datasets, identify anomalies, and provide actionable insights, thereby transforming the internal audit landscape. This article explores how generative AI is being utilized in internal audit functions, supported by real-world examples and scholarly insights.

Enhancing Risk Assessment

Generative AI significantly improves risk assessment by analyzing large volumes of data to detect patterns and anomalies that may indicate potential risks. Traditional audit processes often rely on sampling methods, which can miss critical issues. In contrast, AI can review entire datasets, ensuring a comprehensive assessment.

For example, Deloitte has developed an AI-powered audit tool called "Argus" that enhances risk assessment by analyzing transactional data and identifying unusual patterns indicative of fraud or errors. Argus uses machine learning algorithms to continuously improve its accuracy, providing auditors with a powerful tool to detect risks more effectively (Deloitte, 2023).

Compliance Monitoring

Generative AI can automate compliance monitoring by continuously analyzing transactions and business processes to ensure adherence to regulatory requirements. This capability is particularly valuable in the banking sector, where regulatory environments are complex and constantly evolving.

KPMG's Clara platform is a notable example of AI in compliance monitoring. Clara uses AI to monitor financial transactions and flag potential non-compliance issues in real time. This system helps banks maintain regulatory compliance more efficiently and accurately, reducing the risk of regulatory breaches (KPMG, 2023). According to a report by PwC, banks that implement AI for compliance monitoring can reduce compliance costs by up to 30% (PwC, 2023).

Operational Efficiency

Generative AI enhances operational efficiency in internal audit functions by automating routine tasks such as data collection, analysis, and reporting. This automation allows auditors to focus on more complex and strategic aspects of the audit process.

EY's Helix platform exemplifies the use of AI to enhance operational efficiency in internal audits. Helix automates data extraction and analysis, enabling auditors to quickly identify key issues and trends. This capability not only saves time but also improves the accuracy and depth of audit findings (EY, 2023).

Real-Time Example: JP Morgan Chase's AI-Driven Internal Audits

JP Morgan Chase has implemented AI to transform its internal audit processes. The bank uses AI to analyze transactional data, assess compliance, and identify potential risks. This AI-driven approach has enabled JP Morgan Chase to conduct more thorough and timely audits, significantly improving its internal audit capabilities (JP Morgan Chase, 2023).

According to McKinsey & Company, organizations using AI in internal audits report a 40% increase in audit efficiency and a 25% reduction in audit costs (McKinsey & Company, 2023). These metrics highlight the substantial benefits of integrating AI into internal audit functions.

Predictive Analytics

Predictive analytics powered by generative AI allows internal auditors to anticipate future risks and trends. By analyzing historical data, AI can forecast potential issues, enabling auditors to proactively address risks before they materialize.

An example of predictive analytics in internal auditing is Citibank's use of AI to forecast risk trends and audit outcomes. Citibank's AI system analyzes historical audit data to predict areas of high risk, allowing auditors to prioritize their efforts and focus on the most critical areas (Citibank, 2023). This predictive capability enhances the effectiveness and strategic value of internal audits.

While generative AI offers numerous benefits, its implementation in internal audit functions also presents challenges. Ensuring data privacy and security, managing AI biases, and maintaining regulatory compliance are critical considerations. Banks must invest in robust data governance frameworks and continuously monitor AI systems to mitigate these risks.

Generative AI is redefining internal audit functions in the banking sector by enhancing risk assessment, compliance monitoring, and operational efficiency. Real-world examples from Deloitte, KPMG, EY, JP Morgan Chase, and Citibank demonstrate the transformative potential of AI in internal audits. As banks continue to adopt and integrate AI technologies, they can expect significant improvements in audit effectiveness and efficiency, ensuring they meet the evolving regulatory requirements and maintain robust risk management practices.

Redefining Customer Support and Engagement for Banks with Generative AI Through Cloud Platforms

The banking industry is undergoing a significant transformation, driven by the integration of generative AI and cloud platforms. These technologies are redefining customer support and engagement, enhancing efficiency, personalization, and overall customer satisfaction. This article explores how generative AI, enabled by cloud platforms, is revolutionizing customer support and engagement in the banking sector.

Enhancing Customer Support with Generative AI

Generative AI offers advanced capabilities for improving customer support. By leveraging AI-powered chatbots and virtual assistants, banks can provide instant, accurate, and personalized responses to customer inquiries. These AI-driven systems can handle a wide range of tasks, from answering frequently asked questions to resolving complex issues.

For example, Bank of America's virtual assistant, Erica, uses generative AI to assist customers with their banking needs. Erica can provide balance information, transaction details, and even financial advice based on the user's spending patterns (Bank of America, 2023). This AI-driven support not only improves response times but also enhances the overall customer experience by providing personalized and relevant information.

Personalizing Customer Engagement

Generative AI enables banks to deliver highly personalized customer experiences by analyzing vast amounts of customer data. AI algorithms can identify individual preferences, behaviors, and needs, allowing banks to tailor their services and communications accordingly.

Wells Fargo, for instance, utilizes AI to personalize customer interactions. Their AI system analyzes customer data to offer tailored product recommendations and personalized financial advice (Forrester Research, 2023). This personalized approach helps in building stronger customer relationships and increasing customer loyalty.

Cloud Platforms as Enablers

Cloud platforms play a crucial role in enabling the deployment and scalability of generative AI solutions. By leveraging cloud infrastructure, banks can process large datasets, deploy AI models, and scale their AI-driven services efficiently.

Amazon Web Services (AWS), Microsoft Azure, and Google Cloud are leading cloud platforms that provide the necessary infrastructure for AI applications. For example, HSBC uses Google Cloud to run its AI-driven customer support systems, enabling real-time data processing and analysis (Google Cloud, 2023). The scalability and flexibility offered by cloud platforms are essential for handling the dynamic needs of customer support and engagement in the banking sector.

Real-Time Example: JP Morgan Chase's AI-Driven Customer Support

JP Morgan Chase has integrated AI and cloud technologies to enhance its customer support capabilities. The bank's AI-driven system uses natural language processing (NLP) to understand customer queries and provide accurate responses. This system is hosted on a cloud platform, allowing for real-time processing and scalability.

According to a report by McKinsey & Company, JP Morgan Chase's AI system has significantly improved customer satisfaction by reducing response times and increasing the accuracy of responses (McKinsey & Company, 2023). The use of cloud infrastructure has enabled the bank to handle high volumes of customer interactions efficiently.

Predictive Analytics for Proactive Engagement

Generative AI also enables predictive analytics, allowing banks to anticipate customer needs and engage proactively. By analyzing historical data, AI can predict future behaviors and suggest relevant products and services.

Citibank's AI system, for example, uses predictive analytics to identify customers who might be interested in new financial products based on their transaction history and financial behavior. This proactive approach helps in engaging customers more effectively and enhancing their overall experience (Citibank, 2023).

Challenges and Considerations

While generative AI and cloud platforms offer numerous benefits, their implementation also presents challenges. Ensuring data privacy and security, managing AI biases, and maintaining regulatory compliance are critical issues that banks must address. Robust data governance frameworks and continuous monitoring of AI systems are essential to mitigate these risks.

Generative AI and cloud platforms are revolutionizing customer support and engagement in the banking sector. By enhancing customer support, personalizing engagement, and enabling predictive analytics, these technologies provide significant benefits in terms of efficiency, personalization, and customer satisfaction. Real-world examples from Bank of America, Wells Fargo, HSBC, JP Morgan Chase, and Citibank illustrate the transformative potential of AI and cloud technologies. As banks continue to adopt and integrate these technologies, they can expect to see significant improvements in customer support and engagement, ensuring they meet the evolving needs of their customers and stay competitive in the financial landscape.

4.5. Designing Customer Support Chatbot

Building a customer support chatbot that uses large language models (LLMs) to answer customer queries involves several key steps. Here's a detailed guide to help you through the process:

Step 1: Define the Chatbot's Scope and Objectives

Before you start building the chatbot, clearly define its purpose, the types of queries it will handle, and the scope of its capabilities. Consider whether the chatbot will handle simple FAQs, complex customer support issues, or both.

Step 2: Choose the Right LLM and Platform

Select an appropriate large language model and platform for your chatbot. Popular LLMs include OpenAI's GPT-4, Google's BERT, and Microsoft's Turing. Each has its strengths and can be integrated into various platforms:

- **OpenAI's GPT-4**: Known for its powerful natural language understanding and generation capabilities

- **Google's BERT**: Effective for understanding the context of words in search queries

- **Microsoft's Turing**: Excellent for complex natural language processing tasks

Step 3: Data Preparation

Prepare a dataset that includes the types of queries your chatbot will need to handle. This could include

- **FAQs**: Common questions and their answers
- **Customer Interaction Logs**: Historical data from customer support interactions
- **Product Information**: Detailed information about products or services

Ensure the data is clean, well-structured, and annotated if necessary.

Step 4: Model Training

If using a pre-trained LLM, you can fine-tune it on your specific dataset to improve its performance on your queries. This involves

1. **Preprocessing the Data**: Tokenize text data and convert it into a format suitable for training.
2. **Fine-Tuning the Model**: Train the pre-trained LLM on your dataset to customize it for your specific needs. This step might require substantial computational resources and time.
3. **Testing and Validation**: Test the fine-tuned model to ensure it accurately answers queries. Use a validation dataset to evaluate its performance.

Step 5: Integration

Integrate the LLM into your chatbot framework. This involves

- **Backend Development**: Set up the server and backend systems to handle requests to the LLM.
- **API Integration**: Use APIs to connect the chatbot interface with the LLM. Platforms like OpenAI provide APIs that make it easier to integrate their models.
- **User Interface**: Develop a user-friendly interface for customers to interact with the chatbot. This can be a web-based chat interface, a mobile app, or integration with messaging platforms like Slack or WhatsApp.

Step 6: Building Conversational Flows

Design the conversational flows and logic for the chatbot. This includes

- **Intent Recognition:** Use natural language processing to understand user intents.

- **Response Generation:** Generate appropriate responses based on the user's query and context.

- **Fallback Mechanisms:** Implement fallback mechanisms for queries the chatbot cannot handle, such as transferring to a human agent.

Step 7: Testing and Deployment

Thoroughly test the chatbot in a controlled environment before deploying it. Consider the following testing methods:

- **Unit Testing:** Test individual components of the chatbot

- **Integration Testing:** Ensure that the chatbot works seamlessly with other systems

- **User Acceptance Testing (UAT):** Involve a group of users to test the chatbot and provide feedback

Once testing is complete, deploy the chatbot to a live environment.

Step 8: Monitoring and Maintenance

After deployment, continuously monitor the chatbot's performance and make necessary adjustments. Use analytics to track metrics such as response accuracy, user satisfaction, and engagement rates.

- **Feedback Loop:** Implement a feedback loop where users can rate responses and provide feedback.

- **Regular Updates:** Regularly update the model and the chatbot's knowledge base to keep it relevant and accurate.

Example: Building a Customer Support Chatbot with GPT-4

Tools and Technologies:

- **Backend:** Python with Flask/Django

- **Frontend:** JavaScript with React

- **LLM:** OpenAI GPT-4 via OpenAI API

CHAPTER 4   TRANSFORMING BANKING: THE NEXT FRONTIER

- **Database**: PostgreSQL for storing customer interactions and logs
- **Hosting**: AWS or Google Cloud Platform

Steps:

1. **Set Up Environment**: Configure the development environment with necessary libraries and tools.
2. **Data Preparation**: Collect and preprocess customer support data.
3. **API Integration**: Integrate OpenAI's GPT-4 using OpenAI's API to handle customer queries.
4. **Develop Frontend**: Build a responsive web interface using React.
5. **Testing**: Conduct extensive testing to ensure robustness.
6. **Deployment**: Deploy the chatbot on a scalable cloud platform.
7. **Monitoring**: Set up monitoring tools to track chatbot performance and user interactions

To test the Python code for building a customer support chatbot using OpenAI's GPT-3.5 API, you'll need to ensure that you have the OpenAI API key and the "openai" Python library installed. Here's a complete step-by-step guide to help you run the test:

Step 1: Set Up the Environment

1. Install the "openai" library:

    ```bash
    pip install openai
    ```

2. Obtain an OpenAI API Key: If you don't have an API key yet, you can get one by signing up on the [OpenAI website](https://www.openai.com/).

Step 2: Write the Python Code
Copy and save the following Python code into a file named "test_chatbot.py". Replace 'YOUR_OPENAI_API_KEY' with your actual OpenAI API key.

```python
import openai

# Replace with your OpenAI API key
openai.api_key = 'YOUR_OPENAI_API_KEY'

def get_chatbot_response(prompt):
    response = openai.Completion.create(
        engine="text-davinci-003",  # or "gpt-3.5-turbo" if available
        prompt=prompt,
        max_tokens=150,
        n=1,
        stop=None,
        temperature=0.7
    )
    return response.choices[0].text.strip()

def test_chatbot():
    test_cases = {
        "What are your business hours?": "Our business hours are Monday to Friday, 9 AM to 5 PM.",
        "How can I reset my password?": "You can reset your password by clicking on the 'Forgot Password' link on the login page.",
        "Where are you located?": "We are located at 1234 Main Street, Anytown, USA."
    }

    for question, expected_answer in test_cases.items():
        print(f"Testing question: {question}")
        response = get_chatbot_response(question)
        print(f"Chatbot response: {response}")
        assert expected_answer in response, f"Expected '{expected_answer}' but got '{response}'"

if __name__ == "__main__":
    test_chatbot()
```

Step 3: Run the Python Code

Run the Python script from the command line:

```
python test_chatbot.py
```

- get_chatbot_response Function: This function sends a prompt to the OpenAI API and returns the response from the GPT-3.5 model.

- test_chatbot Function: This function defines a set of test cases and checks if the chatbot's response matches the expected answers. It uses assertions to validate the responses.

- Main Execution: The script runs the "test_chatbot" function if the script is executed as the main program.

Example Output

When you run the script, you should see output similar to this:

```plaintext
Testing question: What are your business hours?
Chatbot response: Our business hours are Monday to Friday, 9 AM to 5 PM.
Testing question: How can I reset my password?
Chatbot response: You can reset your password by clicking on the 'Forgot Password' link on the login page.
Testing question: Where are you located?
Chatbot response: We are located at 1234 Main Street, Anytown, USA.
```

Notes

- Ensure your Internet connection is active as the script makes API calls to OpenAI's servers.

- The "max_tokens" parameter controls the length of the response. Adjust it based on your requirements.

- The "temperature" parameter controls the creativity of the responses. A lower value makes the output more focused and deterministic, while a higher value increases creativity and variation.

By following these steps, you can build and test a customer support chatbot using OpenAI's GPT-3.5 API, ensuring it meets your requirements for accuracy and relevance in answering customer queries.

Revolutionizing Credit Assessment and Lending with Generative AI

The integration of generative AI in credit assessment and lending processes is revolutionizing the banking industry. Traditional credit assessment methods often rely on historical data and predefined criteria, which can be rigid and exclusionary. In contrast, generative AI offers a dynamic and comprehensive approach, analyzing vast amounts of data to generate nuanced credit profiles and lending decisions. This article explores the impact of generative AI on credit assessment and lending, highlighting real-time examples and scholarly insights.

Traditional Credit Assessment Challenges

Traditional credit assessment methods primarily rely on credit scores, financial history, and static criteria to determine an individual's creditworthiness. While these methods have been effective to an extent, they often fail to capture the complete financial picture of potential borrowers. They are also prone to biases and can exclude individuals without extensive credit histories, such as young adults or those new to a country.

According to a study by McKinsey & Company, traditional credit scoring models can lead to inaccurate risk assessments and missed opportunities for lenders to engage with potentially creditworthy customers (McKinsey & Company, 2023). These models typically do not account for non-traditional data sources, which can provide valuable insights into a borrower's ability to repay a loan.

Generative AI in Credit Assessment

Generative AI addresses these limitations by leveraging machine learning algorithms to analyze diverse data sources. These data sources can include transaction histories, social media activity, employment records, and even utility payments. By incorporating a wider range of data, generative AI can create more accurate and comprehensive credit profiles.

One prominent example is Upstart, an AI lending platform that uses generative AI to assess creditworthiness. Upstart's AI models consider over 1,000 data points, including education, employment history, and online behavior, to predict a borrower's ability to repay a loan (Upstart, 2023). This approach has led to a significant reduction in default rates and has allowed the platform to approve more loans, including to individuals with limited credit histories.

According to a report by the National Bureau of Economic Research, AI-driven credit assessments can reduce default rates by up to 50% compared to traditional methods (NBER, 2023). This is achieved by identifying risk factors that conventional models might overlook and by offering more personalized loan terms.

Benefits for Lenders and Borrowers

For lenders, generative AI offers several benefits. It enhances risk management by providing more accurate predictions of borrower behavior, thus reducing the likelihood of defaults. It also enables lenders to expand their customer base by identifying creditworthy individuals who might be excluded by traditional models.

For borrowers, AI-driven credit assessments can lead to fairer and more inclusive lending practices. Individuals with non-traditional credit histories can access loans that were previously out of reach, often at more favorable terms. Additionally, the use of AI can streamline the loan approval process, making it faster and more efficient.

Enhanced Fraud Detection and Security Protocols with Generative AI

In addition to revolutionizing credit assessment, generative AI is also transforming fraud detection and security protocols in the banking sector. Traditional fraud detection systems often rely on predefined rules and historical fraud patterns, which can be limited in their ability to detect new or evolving threats. Generative AI, however, can analyze vast amounts of data in real time, identifying anomalies and potential threats with greater accuracy and speed.

Traditional Fraud Detection Limitations

Traditional fraud detection methods typically involve rule-based systems that flag transactions based on known fraud patterns. While effective to some extent, these systems can generate a high number of false positives, causing inconvenience to customers and additional workload for bank staff. Traditional fraud detection methods, which often rely on rule-based systems, can be rigid and generate a significant number of false positives. This not only inconveniences customers but also places an additional burden on financial institutions. Large Language Models (LLMs) are transforming this landscape by offering more sophisticated and adaptive approaches to detecting fraudulent activities.

LLMs excel in analyzing vast amounts of structured and unstructured data, allowing them to identify complex patterns and subtle anomalies that traditional methods might miss. For instance, research shows that LLMs can enhance fraud detection by analyzing transaction data in real time, considering not only the transaction itself but also the context—such as user behavior and historical data—thereby reducing false positives significantly (Bakumenko et al., 2024; Aisera, 2023; Bank Automation News, 2023).

Furthermore, studies have demonstrated that LLMs can be fine-tuned for specific financial tasks, such as anomaly detection in financial records. By leveraging embeddings from LLMs, financial institutions can improve the accuracy of their fraud detection systems, making it easier to detect irregularities in large datasets with heterogeneous features (Bakumenko et al., 2024).

In practical applications, major financial institutions like JPMorgan are already integrating LLMs into their cybersecurity infrastructure. This allows them to proactively identify and mitigate cyber threats, reducing potential losses and enhancing the overall security of their systems (Bank Automation News, 2023).

These advancements underscore the potential of LLMs to not only improve the accuracy of fraud detection but also to streamline financial operations by reducing the burden of false positives on both customers and bank staff.

According to a study by PwC, traditional fraud detection systems miss up to 30% of fraud cases, particularly those involving sophisticated schemes that do not fit established patterns (PwC, 2023). Additionally, these systems often struggle to adapt to new fraud tactics, leaving banks vulnerable to emerging threats.

Generative AI in Fraud Detection

Generative AI enhances fraud detection by using machine learning algorithms to analyze transaction data, customer behavior, and external data sources. AI models can identify subtle patterns and anomalies that indicate fraudulent activity, even if they do not match previously known patterns.

HSBC's use of AI for fraud detection is a notable example. The bank's AI-driven system monitors transactions in real time, identifying suspicious activities and potential fraud. This system has significantly reduced the number of false positives and improved the accuracy of fraud detection (HSBC, 2023). According to HSBC, the implementation of AI has led to a 20% reduction in fraud-related losses and a 30% increase in fraud detection rates.

Real-Time Analysis and Adaptability

One of the key advantages of generative AI in fraud detection is its ability to analyze data in real time. This capability allows banks to respond to potential threats immediately, minimizing the impact of fraud. Additionally, AI systems can continuously learn and adapt to new fraud tactics, ensuring that they remain effective even as fraudsters develop more sophisticated methods.

JP Morgan Chase's AI-driven fraud detection system exemplifies this approach. The bank uses AI to monitor billions of transactions annually, identifying potential fraud in real time and adapting to new fraud patterns as they emerge (JP Morgan Chase, 2023). This proactive approach has significantly enhanced the bank's security protocols, providing better protection for customers and reducing financial losses.

Benefits for Banks and Customers

For banks, the use of generative AI in fraud detection enhances security and reduces financial losses. By identifying and responding to threats more effectively, banks can protect their assets and maintain customer trust. Additionally, AI-driven systems can reduce the workload for fraud detection teams, allowing them to focus on more complex cases.

For customers, AI-driven fraud detection systems offer better protection against financial crimes. Real-time monitoring and response capabilities can prevent fraudulent transactions before they cause significant harm. Additionally, the reduction in false positives means that legitimate transactions are less likely to be disrupted, improving the overall customer experience.

Generative AI is revolutionizing both credit assessment and fraud detection in the banking sector. By leveraging advanced machine learning algorithms, AI can analyze vast amounts of data to provide more accurate credit profiles and identify potential fraud with greater precision. Real-world examples from Upstart, HSBC, and JP Morgan Chase demonstrate the transformative potential of AI in these critical areas. As banks continue to adopt and integrate AI technologies, they can expect significant improvements in risk management, security, and customer satisfaction.

4.6. Summary

1. Integrating generative AI into existing BFSI infrastructure presents significant challenges due to legacy systems and the complexity of ensuring seamless interoperability.

2. The accuracy and reliability of AI models can be affected by biased or incomplete training data, leading to potential inaccuracies in decision-making processes.

3. The high computational cost and resource requirements for deploying generative AI at scale can be prohibitive for many financial institutions, particularly smaller ones.

4. Ensuring data privacy and security while using AI models that require large datasets poses significant compliance and ethical concerns.

5. The lack of transparency in AI decision-making processes, often referred to as the "black box" problem, complicates the ability to explain and justify AI-driven decisions to stakeholders and regulators.

6. Rapid advancements in AI technologies outpace regulatory frameworks, creating a challenge for institutions to stay compliant with evolving laws and standards.

7. There is a growing risk of AI-driven fraud and cyber threats as AI systems become more sophisticated, requiring continuous adaptation of security protocols.

8. The integration of AI in financial services can lead to job displacement, necessitating the reskilling of the workforce to manage and interact with AI systems.

9. The challenge of maintaining customer trust in AI-driven financial services is significant, especially when dealing with sensitive financial decisions.

10. Finally, the ethical implications of AI, including issues related to fairness, accountability, and the potential for misuse, require ongoing attention and governance from both institutions and regulators.

CHAPTER 5

Innovations in Investment Banking

The financial industry, particularly investment banking and trading, has traditionally been a complex and dynamic sector, characterized by intricate processes, high-stakes decision-making, and significant economic impact. This sector plays a crucial role in the global economy, facilitating capital flow, enabling investments, and driving market activities. However, it also faces numerous challenges, including market volatility, regulatory pressures, and the need for continuous innovation to meet the evolving demands of clients and stakeholders.

5.1. Setting the Scene

Investment banking and trading have long relied on a combination of human expertise, advanced mathematical models, and sophisticated technologies to navigate the complexities of financial markets. Traditionally, these sectors have been driven by data-intensive processes, where traders and analysts sift through vast amounts of information to make informed decisions. The challenges inherent in these activities include managing large datasets, mitigating risks, and optimizing strategies to maximize returns.

Artificial intelligence (AI) has emerged as a transformative force across various industries, offering new capabilities that were previously unattainable. AI's evolution, from rule-based systems to advanced machine learning and deep learning models, has paved the way for its integration into multiple sectors. In finance, AI is now being utilized to enhance efficiency, accuracy, and personalization in service delivery, marking a significant shift from traditional methods.

© Anshul Saxena, Shalaka Verma, Jayant Mahajan 2024
A. Saxena et al., *Generative AI in Banking Financial Services and Insurance*,
https://doi.org/10.1007/979-8-8688-0559-2_5

CHAPTER 5   INNOVATIONS IN INVESTMENT BANKING

The Catalyst for Change

The need for innovation in investment banking and trading is driven by several factors. Market volatility, which can lead to unpredictable price movements and economic instability, requires robust strategies to manage risks and capitalize on opportunities. Additionally, there is a growing demand for personalized financial services, as clients increasingly expect tailored advice and solutions that align with their unique financial goals and risk profiles. Furthermore, higher operational efficiency is essential for maintaining competitiveness and profitability in a rapidly changing market landscape.

AI is positioned as the catalyst for this change, offering capabilities that extend beyond the reach of traditional technologies. AI-driven systems can analyze vast amounts of data at unprecedented speeds, providing insights that enhance decision-making processes. These systems support automation of routine tasks, allowing financial professionals to focus on more strategic activities. Moreover, AI enables more precise risk management by identifying patterns and anomalies that might be missed by human analysts, thus enhancing the overall stability and resilience of financial institutions.

- **Risk Management**: LLMs enable investment banks and trading firms to analyze large datasets rapidly, identifying trends and predicting market movements with greater accuracy. This helps in formulating robust strategies to manage volatility and mitigate risks.

- **Personalized Financial Services**: LLMs contribute to the creation of highly personalized investment strategies by processing client data and preferences. This allows financial advisors to offer tailored advice that aligns with individual financial goals and risk tolerance.

- **Enhanced Decision-Making**: By synthesizing information from diverse sources such as financial news, reports, and market data, LLMs provide traders and investment bankers with actionable insights, leading to more informed decision-making.

- **Algorithmic Trading**: LLMs are used to enhance algorithmic trading strategies by analyzing historical trading patterns, real-time market data, and even sentiment analysis from news and social media, thus optimizing trading performance.

- **Operational Efficiency**: LLMs automate routine tasks such as data entry, report generation, and compliance monitoring, freeing up resources and improving overall operational efficiency within investment banking operations.

- **Market Sentiment Analysis**: LLMs can analyze market sentiment by processing unstructured data from news articles, analyst reports, and social media, helping traders anticipate market reactions and adjust their strategies accordingly.

- **Regulatory Compliance**: LLMs assist in ensuring that investment banks comply with regulatory requirements by automating the monitoring of transactions, detecting suspicious activities, and generating compliance reports.

- **Client Communication**: LLMs improve client communication by powering chatbots and virtual assistants that can handle inquiries, provide real-time updates on market conditions, and offer personalized financial advice based on client data.

- **Portfolio Optimization**: LLMs contribute to portfolio management by analyzing risk factors, asset correlations, and market forecasts, enabling the creation of optimized portfolios that maximize returns while minimizing risk.

- **Innovation in Financial Products**: By analyzing market needs and emerging trends, LLMs help investment banks innovate new financial products that cater to evolving client demands and market conditions.

The integration of AI in investment banking and trading is not just about adopting new technologies; it represents a fundamental shift in how these sectors operate. AI's ability to provide real-time analysis, predictive analytics, and personalized services is reshaping the financial landscape, making it more responsive, efficient, and client-centric. This transformation is setting the stage for a new era in finance, where AI-driven innovations are poised to revolutionize every facet of investment banking and trading, from portfolio management and customer service to trading strategies and risk mitigation.

5.2. AI-Driven Innovations in Finance

Artificial intelligence (AI) encompasses a broad range of technologies, each offering unique capabilities that are highly relevant to financial services. Key AI technologies include machine learning, natural language processing (NLP), and predictive analytics. These technologies have the potential to revolutionize various facets of the financial industry, driving significant improvements in efficiency, accuracy, and client satisfaction.

Highlighting Early Adopters and Success Stories
Several financial institutions have been early adopters of AI, leveraging these technologies to achieve significant advancements.

> **JPMorgan Chase**: The bank has implemented a machine learning-based program called COiN (Contract Intelligence), which processes and interprets legal documents. This has drastically reduced the time required to review documents, improving operational efficiency and accuracy.
>
> **Bank of America**: Bank of America's virtual assistant, Erica, uses NLP and machine learning to provide customers with financial advice, track spending, and offer insights into their financial health. Erica has improved customer satisfaction by providing timely and personalized support.
>
> **BlackRock**: The world's largest asset manager, BlackRock, uses AI-driven tools for portfolio management and risk assessment. Their Aladdin platform leverages machine learning to analyze vast datasets, optimize asset allocation, and manage risk, resulting in enhanced investment strategies and better returns for clients.

Transformative Potential Unveiled
The transformative potential of AI in investment banking and trading is vast, impacting various areas including portfolio management, customer service, trading strategies, fraud detection, and compliance.

Portfolio Management
AI enhances portfolio management by providing personalized investment strategies and optimizing asset allocation. Machine learning models can analyze clients' financial data, market conditions, and historical performance to create tailored investment plans that maximize returns and manage risks.

Example: Wealthfront Wealthfront, a leading robo-advisor, uses AI to offer personalized investment advice and automate portfolio management. The platform analyzes user data to recommend optimal asset allocations and rebalances portfolios automatically based on market changes.

Customer Service

AI-powered chatbots and virtual assistants are revolutionizing customer service in finance. These tools provide instant, 24/7 support, handling routine inquiries, processing transactions, and offering financial advice, thus improving customer satisfaction and engagement.

Example: Erica by Bank of America Bank of America's Erica assists customers with various tasks, from checking balances and tracking spending to making payments and providing personalized financial tips. Erica's success has shown how AI can enhance customer service by making it more responsive and personalized.

Trading Strategies

AI plays a crucial role in developing advanced trading strategies. High-frequency trading (HFT) systems use AI algorithms to execute trades at lightning speed, while predictive analytics help traders forecast market trends and make informed decisions.

Example: Renaissance Technologies

Renaissance Technologies employs AI and machine learning in its trading strategies, analyzing vast amounts of data to identify profitable trading opportunities. The firm's AI-driven approach has consistently delivered high returns, highlighting the potential of AI in trading.

Fraud Detection and Risk Mitigation

AI improves fraud detection and risk mitigation through deep learning and anomaly detection algorithms. These systems analyze transactions in real-time, identifying suspicious activities and assessing risks to protect assets and ensure financial stability.

Example: PayPal

PayPal uses deep learning algorithms to detect fraudulent transactions by analyzing transaction patterns and user behavior. This proactive approach has significantly reduced fraud on the platform, ensuring safer transactions for users.

Compliance

AI streamlines compliance processes by automating the monitoring of transactions, generating compliance reports, and ensuring adherence to regulations. This reduces the regulatory burden on financial institutions and ensures timely compliance with global standards.

Example: Ayasdi

Ayasdi offers AI-driven RegTech solutions that help financial institutions automate compliance processes, detect anomalies, and ensure adherence to regulations. This enhances efficiency and reduces the costs associated with manual compliance checks.

5.3. Preparing for a Revolutionized Future

The ongoing shift towards AI-driven financial services is transforming the landscape of investment banking and trading, with profound implications for professionals, clients, and the market as a whole. As AI technologies continue to evolve and integrate into financial systems, the need for a balanced approach becomes increasingly critical. This involves blending AI capabilities with human expertise to harness the full potential of these innovations while mitigating associated risks.

Impact of AI on Stakeholders

The integration of artificial intelligence (AI) into the financial services industry, particularly in investment banking and trading, has profound implications across multiple facets of the sector. As AI technologies become more advanced, they offer both opportunities and challenges for professionals, clients, and the broader market. AI's ability to automate routine tasks, provide advanced analytics, and generate data-driven insights allows financial professionals to focus on more complex and strategic activities. However, this technological shift also necessitates continuous learning and adaptation, as professionals must acquire new skills to effectively collaborate with AI systems. For clients, AI enhances the personalization and efficiency of financial services but raises concerns about data privacy and the reliability of AI-driven advice. The market itself benefits from improved efficiency and risk management, yet faces potential challenges related to market manipulation and systemic risks. These implications underscore the need for a balanced approach that leverages AI's strengths while addressing its associated risks, ensuring the technology's responsible integration into the financial ecosystem.

Implications for Professionals

For professionals in investment banking and trading, AI offers both opportunities and challenges. On one hand, AI can enhance productivity by automating routine tasks, providing advanced analytical tools, and offering new insights through data analysis. This allows financial professionals to focus on more strategic and complex aspects of their work. On the other hand, there is a need for continuous learning and adaptation to keep pace with technological advancements. Professionals must develop new skills to work effectively alongside AI systems, ensuring they can interpret AI-driven insights and make informed decisions.

Implications for Clients

Clients stand to benefit significantly from AI-driven financial services through personalized and efficient experiences. AI enables tailored investment strategies, proactive financial advice, and enhanced customer service, all of which contribute to higher client satisfaction and engagement. However, clients also need to be aware of the potential risks, such as data privacy concerns and the reliability of AI-driven advice. Transparency and clear communication from financial institutions are essential to build trust and ensure clients understand how their data is being used and the basis for AI-driven recommendations.

Implications for the Market

The integration of AI into financial markets can lead to greater efficiency, improved risk management, and more dynamic trading strategies. AI's ability to process large volumes of data quickly and accurately helps in identifying market trends and opportunities, thus enhancing market liquidity and stability. However, the widespread adoption of AI also introduces challenges such as the potential for market manipulation and systemic risks. Regulatory frameworks need to evolve to address these challenges, ensuring that the benefits of AI are realized while maintaining market integrity and protecting investors.

The Need for a Balanced Approach

To fully harness the potential of AI in financial services, a balanced approach is necessary. This involves leveraging AI capabilities to enhance decision-making, improve efficiency, and deliver personalized services, while also recognizing the importance of human expertise. Financial professionals bring critical thinking, ethical considerations, and experience that are vital for interpreting AI-driven insights and making sound decisions.

Example: AI-Human Collaboration
Goldman Sachs has implemented AI-driven tools to support its traders, providing real-time data analysis and predictive analytics. However, human traders remain crucial for making final decisions, ensuring that the insights generated by AI are used effectively. This collaboration between AI and human expertise enhances overall performance and risk management.

5.4. The Transformative Potential of AI in Investment Banking and Trading

The financial sector is on the brink of a major transformation due to the integration of artificial intelligence (AI). This technology is not just a tool for automation; it is a driving force in redefining investment banking and trading. AI's ability to analyze vast amounts of data at incredible speeds is revolutionizing how financial strategies are developed, enhancing operational efficiency, and shaping the future of the industry. This introduction examines the various ways AI-driven innovations are impacting investment banking and trading, setting the stage for an in-depth exploration of its transformative potential.

AI's capabilities extend far beyond simple data processing. In investment banking and trading, it enables more sophisticated analysis, providing deeper insights and predictive analytics that were previously unattainable. For example, Large Language Models (LLMs) like OpenAI's ChatGPT are used to process financial reports, news articles, and other relevant data sources to generate real-time insights and forecasts. This allows financial institutions to make more informed decisions and stay ahead of market trends.

Refined Portfolio Management Techniques
AI-powered portfolio management represents one of the most significant advancements in the financial sector. Traditional portfolio management relies heavily on historical data and manual analysis, which can be time-consuming and prone to human error. In contrast, AI algorithms and machine learning models can analyze vast datasets, identify patterns, and make predictions with high accuracy. This allows for the creation of personalized investment strategies that are tailored to individual client needs and risk profiles.

Personalized Investment Strategies

AI algorithms can sift through massive amounts of financial data to identify trends and correlations that may not be immediately apparent to human analysts. For example, an AI system can analyze a client's transaction history, spending habits, and financial goals to create a customized investment portfolio. This level of personalization helps clients achieve their financial objectives more effectively and efficiently.

Optimization of Asset Allocation

One of the key benefits of AI in portfolio management is its ability to optimize asset allocation. Traditional methods of asset allocation often involve a static approach, where investments are distributed based on predefined criteria. AI, however, can dynamically adjust asset allocations based on real-time data and market conditions. This ensures that portfolios are always optimized to maximize returns while minimizing risks.

Case Study: BlackRock BlackRock, the world's largest asset manager, has been a pioneer in incorporating AI into its portfolio management processes. The company uses AI-driven tools to analyze market data, assess risk factors, and optimize asset allocation. By leveraging AI, BlackRock has been able to enhance its investment strategies and deliver better returns for its clients.

Risk Management and Mitigation

AI also plays a crucial role in risk management. Machine learning models can identify potential risks by analyzing historical data and predicting future market movements. This allows portfolio managers to take proactive measures to mitigate risks, such as adjusting asset allocations or implementing hedging strategies. AI's predictive capabilities help in identifying market anomalies and potential downturns, allowing for timely interventions that can protect client investments.

Example: Robo-advisors

Robo-advisors are a prime example of AI-driven portfolio management. These digital platforms use algorithms to manage investment portfolios automatically. They assess an investor's risk tolerance, financial goals, and market conditions to provide personalized investment advice and portfolio management. Robo-advisors have democratized access to sophisticated investment strategies, making them available to a broader audience at a lower cost.

Enhanced Returns

AI's ability to process and analyze large datasets leads to more informed investment decisions, which in turn can enhance returns. By continuously monitoring market conditions and adjusting strategies in real-time, AI ensures that investment portfolios are always aligned with the latest market trends. This dynamic approach helps in capturing new opportunities and maximizing returns.

Case Study: Wealthfront Wealthfront, a leading robo-advisor, uses AI and machine learning to optimize its clients' investment portfolios. The platform analyzes vast amounts of data to provide personalized investment recommendations, adjust asset allocations, and implement tax-efficient strategies. Wealthfront's AI-driven approach has helped it deliver competitive returns while minimizing risks for its clients.

The integration of AI in investment banking and trading is reshaping the industry, offering refined portfolio management techniques that enhance returns and manage risks more efficiently. Through personalized investment strategies, optimized asset allocation, and advanced risk management, AI is setting new standards in the financial sector. As AI technology continues to evolve, its impact on investment banking and trading will only grow, leading to more innovative solutions and improved financial outcomes for clients.

The financial services industry is increasingly leveraging Large Language Models (LLMs) to create AI-enabled customer-centric solutions that significantly enhance customer satisfaction and engagement. These advancements focus on developing tailored banking products, improving customer service, and providing more personalized financial experiences. LLMs enable banks and financial institutions to analyze vast amounts of customer data, including transaction histories, spending patterns, and personal preferences, to develop customized products that align closely with each customer's unique financial goals and lifestyle.

For instance, Capital One employs LLMs to analyze customer data and offer personalized banking products. The bank's AI-driven platform can identify customers who may benefit from specific credit card features or loan products, providing them with more relevant and attractive financial options. This ability to tailor financial products not only improves customer satisfaction but also enhances the bank's ability to meet the diverse needs of its clientele, ultimately driving customer loyalty and retention.

By leveraging LLMs, financial institutions can move beyond generic offerings, instead providing highly personalized and effective financial solutions.

Improved Customer Service Through Chatbots and AI Assistants

AI-powered chatbots and virtual assistants are revolutionizing customer service in the financial sector. These tools provide instant support to customers, handling inquiries, processing transactions, and offering financial advice. By automating routine tasks, chatbots and AI assistants free up human agents to focus on more complex customer issues.

Case Study: Bank of America's Erica Bank of America's AI-powered virtual assistant, Erica, helps customers manage their finances by providing personalized insights, answering questions, and assisting with transactions. Erica's ability to understand and respond to a wide range of customer queries has significantly improved the bank's customer service efficiency and satisfaction.

Impact on Customer Satisfaction and Engagement

AI-driven customer-centric solutions have a profound impact on customer satisfaction and engagement. By offering personalized experiences and immediate support, financial institutions can build stronger relationships with their customers. AI also enables proactive engagement, where banks can anticipate customer needs and offer solutions before issues arise.

Example: Wells Fargo

Wells Fargo uses AI to enhance customer engagement by sending personalized notifications and financial health insights to its customers. This proactive approach helps customers stay informed about their financial status and makes it easier for them to achieve their financial goals.

Navigating Trading and Market Predictions

AI is transforming trading strategies and market predictions, allowing traders to analyze vast datasets, forecast trends with greater accuracy, and execute trades at speeds that surpass human capabilities.

Analyzing Vast Datasets for Insights

AI's ability to process and analyze large volumes of data is invaluable in trading. By examining historical data, market news, social media sentiment, and other relevant information, AI systems can identify patterns and generate actionable insights. This comprehensive analysis enables traders to make more informed decisions.

Example: Renaissance Technologies

Renaissance Technologies, a hedge fund known for its quantitative trading strategies, uses AI and machine learning to analyze vast datasets and identify trading opportunities. The firm's AI-driven approach has consistently delivered impressive returns by uncovering insights that human analysts might miss.

Forecasting Market Trends with Greater Accuracy

AI's predictive capabilities are enhancing market trend forecasts. Machine learning models can analyze historical trends and current market conditions to predict future movements with a high degree of accuracy. This allows traders to anticipate market shifts and adjust their strategies accordingly.

Case Study: Goldman Sachs Goldman Sachs employs AI to improve its market predictions and trading strategies. By analyzing a wide range of data sources, AI models at Goldman Sachs can forecast market trends more accurately, helping traders make better-informed decisions.

Enabling High-Frequency Trading

High-frequency trading (HFT) relies heavily on AI and machine learning to execute trades at lightning speeds. AI algorithms can process and react to market data in milliseconds, executing thousands of trades in a fraction of a second. This speed and precision give traders a significant advantage in fast-moving markets.

Example: Citadel Securities

Citadel Securities is a leading player in high-frequency trading, leveraging AI to execute trades at extremely high speeds. The firm's AI-driven trading systems can process and act on market information faster than any human, enabling it to capitalize on fleeting trading opportunities.

AI's integration into trading and market predictions is not only enhancing the accuracy and efficiency of trading strategies but also opening up new possibilities for innovation in the financial markets. As AI technology continues to advance, its role in shaping the future of trading will become even more prominent, driving further improvements in performance and profitability.

Modern Fraud Detection and Risk Mitigation

Advancements in artificial intelligence (AI) are revolutionizing fraud detection and risk mitigation in the financial sector. AI's ability to analyze vast amounts of data in real-time and identify patterns has significantly enhanced the capabilities of financial institutions to protect assets and ensure stability.

Deep Learning in Fraud Detection

Deep learning algorithms excel at recognizing complex patterns and anomalies within large datasets. By training on historical data, these algorithms can detect fraudulent activities that may be missed by traditional methods. They continually learn and adapt, improving their accuracy over time.

Example: PayPal

PayPal employs deep learning algorithms to detect fraudulent transactions. By analyzing millions of transactions per day, the system identifies unusual patterns that may indicate fraud. This proactive approach has significantly reduced fraudulent activity on the platform, ensuring safer transactions for users.

Anomaly Detection Algorithms

Anomaly detection algorithms are designed to identify deviations from normal behavior, which are often indicative of fraud. These algorithms can analyze various data points, such as transaction amounts, locations, and times, to spot irregularities in real-time.

Case Study: JP Morgan Chase JP Morgan Chase uses AI-powered anomaly detection to monitor its transactions. The system flags any activity that deviates from established patterns, allowing the bank to investigate and respond to potential fraud quickly. This real-time monitoring helps in minimizing losses and maintaining customer trust.

AI-driven real-time risk assessment empowers financial institutions to swiftly respond to emerging threats by continuously analyzing data and assessing potential risks, whether related to market fluctuations or fraudulent activities. For example, Mastercard employs an AI-driven system that evaluates transactions as they occur, using data from multiple sources to determine the likelihood of fraud. This real-time analysis allows for the prevention of fraudulent transactions before they are completed, safeguarding both the company and its customers. By integrating AI in fraud detection and risk mitigation, financial institutions protect their assets and contribute to overall financial stability, reducing the potential for significant financial losses.

5.5. The New Era of Financial Planning and Advisory

AI is transforming financial planning and advisory services, offering personalized financial advice, enhancing wealth management, and redefining client-advisor interactions. These advancements are making financial planning more accessible and effective for a broader range of clients.

Personalized Financial Advice

AI-driven tools can analyze individual financial situations and provide tailored advice. By considering factors such as income, spending habits, and financial goals, AI systems can offer personalized recommendations that are more accurate and relevant than generic advice.

Example: Betterment

Betterment, a robo-advisor, uses AI to provide personalized financial advice to its users. The platform assesses each client's financial situation and goals, creating customized investment plans that align with their needs. This personalized approach helps clients achieve their financial objectives more efficiently.

Robo-advisors

Robo-advisors are digital platforms that use algorithms to manage investment portfolios automatically. They offer low-cost, efficient, and accessible financial planning services, democratizing access to high-quality investment advice.

Case Study: Wealthfront Wealthfront is a leading robo-advisor that uses AI to optimize its clients' investment portfolios. The platform provides personalized investment recommendations, adjusts asset allocations, and implements tax-efficient strategies. This AI-driven approach has made sophisticated financial planning accessible to a wider audience.

Impact on Long-Term Financial Planning and Wealth Management

AI's role in financial planning extends to long-term strategies and wealth management. By continuously analyzing market conditions and individual financial data, AI tools can help clients make informed decisions that align with their long-term goals.

Example: Charles Schwab's Intelligent Portfolios

Charles Schwab's Intelligent Portfolios use AI to provide automated investment management services. The platform continuously monitors and rebalances portfolios to ensure they remain aligned with clients' long-term financial goals. This helps clients stay on track and achieve their desired financial outcomes.

Enhancing Client-Advisor Interactions
AI is also enhancing the interactions between clients and human advisors. By providing advisors with detailed data and insights, AI tools enable more informed and meaningful conversations. Advisors can offer more personalized and strategic advice, improving the overall client experience.

Example: Morgan Stanley's Next Best Action
Morgan Stanley's Next Best Action platform uses AI to provide advisors with tailored recommendations for their clients. By analyzing client data and market conditions, the platform suggests specific actions that advisors can discuss with their clients. This enhances the quality of advice and strengthens client-advisor relationships.

The integration of AI in financial planning and advisory services is not only making financial advice more personalized and accessible but also improving the efficiency and effectiveness of wealth management. As AI technology continues to advance, its impact on financial planning will grow, offering even more innovative solutions for managing personal finances and achieving long-term financial goals.

Complying with the Evolving Regulatory Landscape
The integration of AI in the financial industry brings numerous benefits, but it also presents significant regulatory challenges. Ensuring compliance in an AI-driven environment requires transparent AI models, adherence to global financial regulations, and leveraging AI to streamline compliance processes.

Transparent AI Models
Transparency in AI models is crucial for regulatory compliance. Financial institutions must ensure that their AI systems are explainable, meaning the decision-making processes of these models can be understood and scrutinized. This is important for gaining regulatory approval and maintaining trust with clients.

Example: Explainable AI (XAI) in Banking
Explainable AI (XAI) techniques are being adopted by banks to enhance the transparency of their AI models. For instance, HSBC uses XAI to ensure that their AI-driven credit scoring systems are transparent. This helps in explaining credit decisions to regulators and customers, thereby enhancing trust and compliance.

Adherence to Global Financial Regulations
AI systems in the financial industry must comply with a variety of global financial regulations. These include data protection laws, Anti-money Laundering (AML)

regulations, and financial reporting standards. Financial institutions need to stay updated with these regulations and ensure their AI systems adhere to them.

Case Study: GDPR Compliance The General Data Protection Regulation (GDPR) in Europe sets strict guidelines for data protection and privacy. Financial institutions using AI must ensure that their data processing practices comply with GDPR. This includes obtaining explicit consent from customers, ensuring data accuracy, and implementing measures to protect data privacy.

Streamlining Compliance Processes

AI can also be used to streamline compliance processes, reducing the regulatory burden on financial institutions. AI systems can automate the monitoring of transactions for suspicious activities, generate compliance reports, and ensure that all regulatory requirements are met in real-time.

Example: Regulatory Technology (RegTech)

RegTech solutions use AI to enhance regulatory compliance. Companies like Ayasdi offer AI-driven platforms that help financial institutions automate compliance processes, detect anomalies, and ensure adherence to regulations. This reduces the time and cost associated with manual compliance checks.

Future Prospects: Sculpting the Future of Financial Services

The future of financial services with AI at its core promises ongoing innovations, potential disruptions, and a move towards sustainable, inclusive, and resilient financial ecosystems. This section explores the future landscape, ethical considerations, and the collaborative efforts required to navigate the evolving financial sector.

Ongoing Innovations and Potential Disruptions

AI will continue to drive innovations in financial services, from advanced analytics and personalized banking experiences to automated trading and enhanced risk management. However, these advancements also come with potential disruptions, including job displacement and increased competition from tech-driven financial startups.

Example: FinTech Startups

FinTech startups are leveraging AI to offer innovative financial solutions that challenge traditional banking models. Companies like Stripe and Robinhood use AI to provide seamless payment processing and investment services, respectively, disrupting conventional financial services and pushing the industry towards more customer-centric models.

Fostering Sustainable and Inclusive Financial Ecosystems

AI has the potential to create more sustainable and inclusive financial systems. By providing personalized financial services, AI can help individuals and small businesses access credit and investment opportunities that were previously out of reach. This fosters financial inclusion and supports economic growth.

Case Study: Microfinance and AI AI-powered microfinance platforms, like Kiva, use machine learning to assess creditworthiness and provide loans to underserved populations. This helps in promoting financial inclusion and supporting sustainable development by empowering individuals and small businesses.

Ethical Considerations and Responsible AI Development

As AI becomes more integrated into financial services, ethical considerations must be addressed. This includes ensuring fairness, avoiding bias in AI algorithms, and protecting customer privacy. Responsible AI development involves creating systems that are not only effective but also ethical and trustworthy.

Example: AI Ethics in Finance

Financial institutions are establishing AI ethics boards to oversee the development and deployment of AI systems. These boards ensure that AI technologies are developed responsibly, with considerations for fairness, transparency, and accountability.

Collaborative Efforts for Navigating the Future of Finance

Navigating the future of finance requires collaboration between financial institutions, regulators, technology providers, and other stakeholders. By working together, these entities can develop standards, share best practices, and ensure that AI-driven innovations benefit the entire financial ecosystem.

Example: Industry Collaboration

Initiatives like the Global AI Finance Alliance bring together financial institutions, regulators, and AI experts to address the challenges and opportunities of AI in finance. This collaborative approach helps in developing guidelines and standards that promote the responsible use of AI in the financial sector.

The future of financial services with AI at its core is bright, promising innovations that enhance efficiency, inclusivity, and sustainability. However, achieving this future requires a balanced approach that considers ethical implications, regulatory compliance, and collaborative efforts to navigate the evolving financial landscape.

5.6. Refined Portfolio Management Techniques

Traditional Portfolio Management Strategies and Their Limitations

Traditional portfolio management strategies rely heavily on historical data and manual analysis, which can be time-consuming and prone to human error. These strategies typically use diversification and risk tolerance assessments to allocate assets, but they often lack the agility and precision needed to respond quickly to market changes and individual investor needs.

Emergence of AI in Portfolio Management

AI has revolutionized portfolio management by introducing advanced data analysis and decision-making capabilities. Machine learning algorithms, predictive analytics, and natural language processing (NLP) allow for the processing and analysis of vast amounts of data at high speeds. This leads to more accurate predictions, timely adjustments, and tailored investment strategies that traditional methods cannot achieve.

Personalized Investment Strategies

AI-Driven Customization

AI algorithms analyze detailed investor profiles, including financial goals, risk tolerance, and market conditions, to create customized investment strategies. These algorithms continuously learn from new data, allowing them to adapt and refine strategies in real-time to align with changing market dynamics and investor needs.

Data Analysis for Personalization

AI systems evaluate various factors to personalize investment strategies, including:

- **Investor Profiles:** Personal data, financial history, and investment objectives

- **Market Conditions:** Current market trends, economic indicators, and geopolitical events

- **Risk Tolerance:** Investor's capacity to withstand financial loss based on their financial situation and preferences

Example: Betterment

Betterment uses AI to provide personalized investment advice. The platform collects data on each client's financial situation and goals, then uses this information to recommend an optimal investment strategy. Betterment's AI continuously monitors the portfolio, making automatic adjustments to maintain the desired risk level and optimize returns.

Optimizing Asset Allocation

AI Models for Predictive Analytics and Machine Learning

AI models use predictive analytics and machine learning to identify the optimal asset allocation by balancing expected returns and risks. These models analyze historical data, market conditions, and individual investor profiles to make informed decisions about where to allocate assets.

AI-driven asset allocation is highly dynamic, adjusting in real-time to market changes and individual investor circumstances. This ensures that portfolios remain optimized, even in volatile market conditions.

Example: BlackRock's Aladdin Platform

BlackRock's Aladdin platform uses AI and machine learning to optimize asset allocation. The system continuously analyzes vast datasets, assesses risk, and dynamically adjusts asset allocations based on real-time market conditions. This approach ensures that portfolios are always optimized for maximum returns while managing risks effectively.

Improving Risk Management

AI improves risk management by foreseeing potential market downturns, analyzing historical data for risk patterns, and implementing strategies to mitigate those risks. Machine learning models can predict potential risks and provide actionable insights to prevent significant financial losses.

Example: PayPal

PayPal uses deep learning algorithms to detect fraudulent transactions by analyzing transaction patterns and user behavior. This proactive approach has significantly reduced fraud on the platform, ensuring safer transactions for users.

AI-Driven Risk Mitigation Strategies

AI-driven risk mitigation strategies include continuous monitoring of market conditions, real-time analysis of portfolio performance, and dynamic adjustments to asset allocations. These strategies help in identifying and mitigating risks before they materialize, ensuring financial stability and protecting investor assets.

Example: Wealthfront

Wealthfront's robo-advisor platform uses AI to optimize portfolios and manage risks. The platform implements tax-efficient strategies, automatic rebalancing, and real-time risk assessment to enhance returns while mitigating potential risks. Wealthfront's AI-driven approach ensures that portfolios remain aligned with clients' financial goals and risk tolerance.

CHAPTER 5   INNOVATIONS IN INVESTMENT BANKING

The Role of Big Data and Advanced Analytics

Leveraging Big Data for Deeper Insights

AI leverages big data and advanced analytics to gain deeper insights into market trends, investor behavior, and economic indicators. By analyzing vast datasets, AI can identify correlations and trends that inform more accurate predictions and strategic decisions.

Integration with Other Technologies

AI's effectiveness is further enhanced through integration with other technologies like the Internet of Things (IoT) and blockchain. These technologies enrich data quality and accessibility, providing AI systems with more comprehensive and reliable data sources.

Example: IoT and Blockchain Integration

IoT devices can collect real-time data on various economic activities, while blockchain ensures data integrity and security. Financial institutions can use this enriched data to feed AI models, leading to more informed portfolio management decisions.

Challenges and Considerations

Data Privacy Concerns

Implementing AI in portfolio management involves handling sensitive financial data, raising data privacy concerns. Financial institutions must ensure robust data protection measures to maintain client trust and comply with regulations like GDPR.

Transparent Algorithms

The need for transparent algorithms is critical to gaining regulatory approval and client trust. Explainable AI (XAI) techniques are essential to make AI-driven decisions understandable and auditable.

Example: HSBC and Explainable AI

HSBC uses XAI to ensure their AI-driven credit scoring systems are transparent, making it easier to explain decisions to regulators and customers, thereby enhancing trust and compliance.

Human Oversight

While AI can automate many aspects of portfolio management, human oversight remains essential. Financial professionals need to interpret AI-driven insights and make final decisions, ensuring ethical considerations and strategic alignment.

Future Prospects

Advancements in AI Algorithms

Future advancements in AI-driven portfolio management will likely involve the development of more sophisticated algorithms capable of deeper analysis and more

CHAPTER 5 INNOVATIONS IN INVESTMENT BANKING

accurate predictions. These advancements will further enhance the personalization and optimization of investment strategies.

Integration with New Data Sources

AI's ability to integrate with new data sources, such as alternative data (e.g., social media trends, satellite imagery), will provide even richer insights, leading to more informed decisions.

Broader Impact on Financial Advisory

The broader impact on the financial advisory landscape includes a shift towards more tech-savvy professionals who can leverage AI technologies effectively. Continuous education and adaptation will be essential for financial advisors to remain relevant and provide value in an AI-driven world.

Customer-Centric Financial Solutions

Introduction to AI in Customer-Centric Solutions

Traditional Approach to Customer Service and Product Offerings

Historically, financial services relied on a one-size-fits-all approach to customer service and product offerings. Standardized products such as loans, credit cards, and savings accounts were common, with limited customization based on individual needs. Customer service was primarily managed through in-person interactions, phone calls, and emails, leading to long wait times and inconsistent service quality.

Shift Towards Customer-Centric Solutions

AI has facilitated a significant shift towards customer-centric solutions in financial services, focusing on understanding and addressing unique customer needs and preferences. AI enables financial institutions to analyze vast amounts of data, predict needs, and deliver tailored products and services in real-time.

Customized Savings Plans and Investment Options

AI-driven tools create personalized savings plans and investment options based on individual financial goals and risk tolerance. By continuously analyzing market trends and customer financial behavior, these tools adjust recommendations to maximize returns and align with changing needs.

Example: AI-Generated Investment Portfolios

Platforms like Wealthfront and Betterment use AI to offer personalized investment portfolios. These platforms leverage LLMs to analyze user data and market conditions, generating customized investment strategies. The AI continuously monitors the portfolio, making adjustments to asset allocations in response to market changes, ensuring that the investment strategy remains aligned with the customer's goals.

AI-Enhanced Customer Service

24/7 Customer Support with Chatbots and AI Assistants

AI-powered chatbots and virtual assistants provide instant, 24/7 support, handling routine inquiries, processing transactions, and offering financial advice. By automating these tasks, AI reduces wait times and improves service consistency, leading to higher customer satisfaction.

Example: Erica by Bank of America

Bank of America's virtual assistant, Erica, uses AI to assist customers with various tasks, from checking balances and tracking spending to making payments and providing personalized financial tips. Erica's use of NLP and machine learning enables it to understand and respond to a wide range of customer queries, enhancing the overall customer service experience.

Case Studies Showcasing Effectiveness

Case Study: Swedbank's Nina   Swedbank implemented an AI-powered virtual assistant named Nina to handle customer inquiries. Nina successfully managed over 80% of customer interactions, significantly reducing wait times and improving resolution rates. Customer satisfaction increased as Nina provided quick and accurate responses, demonstrating the effectiveness of AI in enhancing customer service.

Case Study: HSBC's Amy   HSBC introduced Amy, an AI chatbot, to handle customer inquiries related to mortgages. Amy was able to handle thousands of queries simultaneously, providing instant responses and reducing the need for human intervention. This led to a substantial improvement in customer satisfaction and operational efficiency.

Personalized Communication and Marketing

Crafting Personalized Marketing Messages

AI enables financial institutions to craft personalized marketing messages and communication strategies by analyzing customer data and behavior. This targeted approach ensures that customers receive relevant offers and information tailored to their interests and needs.

Example: AI-Driven Marketing at Wells Fargo

Wells Fargo uses AI to analyze customer data and deliver personalized marketing messages. By targeting customers with relevant offers, such as personalized loan products or investment opportunities, the bank enhances customer engagement and conversion rates.

Impact on Customer Engagement and Loyalty

Personalized communication significantly impacts customer engagement and loyalty. Customers are more likely to respond positively to messages that resonate with their individual needs and preferences. AI-driven personalization helps build stronger relationships and fosters long-term loyalty.

Example: Personalized Email Campaigns

AI-powered platforms like Salesforce's Einstein use machine learning to personalize email campaigns. By analyzing customer behavior and preferences, the system crafts tailored messages that improve open rates, click-through rates, and overall engagement.

Improving Customer Engagement Through AI

Encouraging Active Customer Engagement

AI-driven tools and applications encourage active customer engagement by providing interactive and user-friendly platforms for financial planning, budgeting, and investment advice. These tools help customers stay engaged with their finances and make informed decisions.

Example: Mint and Personal Capital

Budgeting apps like Mint and financial planning tools like Personal Capital use AI to provide personalized insights and recommendations. These apps analyze users' financial data to offer tailored budgeting advice, spending analysis, and investment recommendations, encouraging users to actively engage with their finances.

Role of Gamification and Interactive Features

Gamification and interactive features enhance the user experience and foster a deeper connection with financial institutions. By incorporating elements like rewards, challenges, and progress tracking, these features make financial management more engaging and enjoyable.

Example: Monzo and Simple

Banks like Monzo and Simple use gamification to engage customers. Monzo's app includes features like spending targets and real-time notifications, while Simple offers goal-based savings and progress tracking. These interactive features help users stay motivated and engaged with their financial goals.

5.7. Challenges and Ethical Considerations

Privacy Concerns and Data Security

Implementing AI for customer-centric solutions involves handling sensitive financial data, raising privacy and security concerns. Financial institutions must ensure robust data protection measures to maintain client trust and comply with regulations like GDPR.

Risk of Bias in AI Algorithms

AI algorithms can inadvertently perpetuate biases present in training data, leading to unfair outcomes. Financial institutions must develop transparent and fair AI models to avoid bias and ensure equitable treatment of all customers.

Ethical Considerations in Data Use

Using personal data to tailor products and services raises ethical considerations. Financial institutions must be transparent about how they use customer data, ensuring that customers are aware and consent to data usage. Maintaining trust is crucial for long-term customer relationships.

Example: Transparent AI Practices

Companies like FICO are developing explainable AI models that provide clear insights into how decisions are made. This transparency helps build trust and ensures that customers understand the basis for personalized offers and recommendations.

Impact on Customer Satisfaction and Engagement

Positive Impact of AI-Enabled Solutions

Studies and surveys have shown that AI-enabled customer-centric solutions significantly improve customer satisfaction, loyalty, and engagement. Personalized services and responsive customer support contribute to a better overall experience.

Example: JD Power's Banking Satisfaction Study

JD Power's Banking Satisfaction Study found that banks using AI to personalize services and improve customer support saw higher satisfaction scores. Customers appreciated the convenience and relevance of AI-driven interactions, leading to increased loyalty.

Contribution to Growth and Competitive Advantage

Enhanced customer satisfaction and engagement drive growth and provide a competitive advantage for financial institutions. Banks that leverage AI effectively can differentiate themselves in a crowded market, attracting and retaining more customers.

Traditional Practices in Investment Banking and Trading

Investment banking and trading have traditionally relied on human expertise and quantitative models to make decisions. Analysts sift through historical data, market trends, and economic indicators to identify investment opportunities and manage risks. However, these traditional practices face significant limitations:

- **Volume of Data:** The exponential growth of financial data makes it challenging for human analysts to process and extract actionable insights in a timely manner.

- **Speed of Execution:** Manual processes can lead to delays in decision-making and execution, which are critical in fast-moving markets.

- **Risk Management:** Traditional risk assessment methods often struggle to keep up with the rapid changes and complexities of global financial markets.

AI as a Transformative Force

AI introduces capabilities that address these limitations, enabling real-time data processing, predictive analytics, and automation of complex tasks. By leveraging AI technologies, investment banks and trading firms can enhance efficiency, accuracy, and responsiveness, transforming the financial sector.

Revolutionizing Portfolio Management

Enhancing Portfolio Management with AI

AI enhances portfolio management by enabling personalized investment strategies, real-time asset allocation adjustments, and comprehensive risk assessments. These advancements are driven by sophisticated algorithms and machine learning models that continuously learn from new data and market conditions.

Personalized Strategy Development

AI algorithms analyze individual investor profiles, financial goals, risk tolerance, and current market conditions to create tailored investment strategies. This personalized approach helps maximize returns while managing risks according to the specific needs of each investor.

Example: LLM-Powered Robo-advisors

Robo-advisors such as Wealthsimple use Large Language Models (LLMs) to develop personalized investment plans for clients. By analyzing user data, these platforms provide tailored recommendations and automate the investment process, continuously adjusting portfolios to align with clients' financial goals and risk preferences. LLMs help interpret complex financial data and generate personalized insights, making the investment advice more accessible and comprehensible to clients.

Real-Time Asset Allocation Adjustments
AI-driven asset allocation models dynamically adjust investments based on real-time market data. This capability ensures that portfolios remain optimized to capture emerging opportunities and mitigate risks as market conditions change.

Example: LLM-Enhanced UBS Delta
UBS Delta, a portfolio management platform, uses AI and LLMs to provide real-time insights and recommendations on asset allocation. The platform leverages LLMs to process and interpret vast amounts of financial data, enabling portfolio managers to adjust their strategies dynamically. This integration ensures that client portfolios remain aligned with the latest market developments and personalized advice.

Comprehensive Risk Assessment
AI-powered risk assessment tools analyze historical and real-time data to identify potential risks. These tools enable portfolio managers to implement proactive measures to mitigate risks, enhancing portfolio stability and resilience.

Example: Morgan Stanley's Next Best Action with LLMs
Morgan Stanley's Next Best Action platform leverages AI and LLMs to assess risk and provide actionable recommendations. The platform uses LLMs to interpret complex datasets, identifying potential risks and opportunities. This helps advisors make informed decisions to protect and grow client assets, ensuring that risk management strategies are both proactive and personalized.

Impact on Customer Satisfaction and Operational Efficiency
The integration of AI in customer service has led to reduced wait times, higher resolution rates, and increased overall customer satisfaction. By automating routine tasks, AI allows human agents to focus on more complex issues, further improving service quality.

Advancing Trading Strategies

AI in Advanced Trading Strategies
AI facilitates advanced trading strategies by enabling algorithmic trading that can execute orders at optimal prices and predictive analytics that anticipate market movements. These capabilities allow trading firms to achieve superior results compared to traditional methods.

Algorithmic Trading

AI-driven algorithmic trading systems analyze vast amounts of market data in real-time to identify profitable trading opportunities and execute trades at the optimal moment. This reduces human error and increases trading efficiency.

Example: LLM-Enhanced Algorithmic Trading

Case Study: Two Sigma Two Sigma, a quantitative hedge fund, uses AI and LLMs to enhance its trading strategies. The firm's AI-driven system processes massive datasets, including market data, news articles, and social media sentiment, to identify trading opportunities. LLMs help interpret this data and generate actionable insights, leading to more informed and timely trading decisions.

Predictive Analytics

AI's predictive analytics capabilities enable trading firms to anticipate market movements by analyzing historical data, current market conditions, and economic indicators. This allows traders to make more informed decisions and adjust their strategies proactively.

Example: Predictive Trading Systems

Case Study: Kensho by S&P Global

Kensho, an AI platform by S&P Global, uses predictive analytics to forecast market trends. By analyzing diverse datasets, including financial reports, geopolitical events, and social media activity, Kensho's AI-driven system generates accurate predictions about market movements. This has enabled traders to make more strategic decisions and achieve better trading outcomes.

Real-World Scenarios of AI-Powered Trading Systems

Case Study: Sentient Technologies Sentient Technologies has developed an AI-powered trading system that uses deep learning and evolutionary algorithms to identify trading opportunities. The system continuously learns from new data, adapting its strategies to changing market conditions. This approach has delivered superior trading results, demonstrating the potential of AI in enhancing trading performance.

CHAPTER 5   INNOVATIONS IN INVESTMENT BANKING

Case Study: Numerai   Numerai, a hedge fund, leverages AI and LLMs to crowdsource trading algorithms from data scientists worldwide. The firm's AI platform integrates these algorithms to make collective trading decisions, resulting in innovative and effective trading strategies. This collaborative approach has led to impressive trading performance, showcasing the power of AI in advancing trading strategies.

Enhancing Fraud Detection and Security

AI's Role in Bolstering Fraud Detection and Security

AI enhances fraud detection and security measures through advanced pattern recognition, anomaly detection, and predictive models. These capabilities enable financial institutions to preemptively identify fraudulent activities and strengthen their security frameworks.

Pattern Recognition and Anomaly Detection

AI systems can analyze vast amounts of transactional data to identify patterns and detect anomalies that may indicate fraudulent activities. By learning from historical data, AI models can recognize unusual behavior that deviates from normal patterns.

Example: LLM-Powered Fraud Detection

Case Study: PayPal

PayPal uses AI and LLMs to enhance its fraud detection capabilities. The system analyzes transaction patterns and user behavior to detect anomalies and flag suspicious activities. LLMs help interpret the data, providing detailed explanations and actionable insights, enabling swift and effective responses to potential fraud.

Predictive Models for Fraud Prevention

AI-driven predictive models can assess the likelihood of fraudulent activities based on historical and real-time data. These models enable financial institutions to implement preventive measures and mitigate risks before fraud occurs.

Case Study: Mastercard's Decision Intelligence   Mastercard's Decision Intelligence platform uses AI to predict and prevent fraud. By analyzing transaction data and user behavior, the platform's AI-driven models assess the risk of each transaction in real-time, providing recommendations to approve or decline transactions. This proactive approach has significantly reduced fraud rates and enhanced security.

Case Study: Capital One Capital One employs AI to detect and prevent fraud across its financial services. The AI system analyzes millions of transactions daily, identifying suspicious patterns and flagging potential fraud. This has improved the bank's ability to prevent fraud and protect its customers.

Case Study: HSBC HSBC uses AI and LLMs to monitor transactions for fraudulent activities. The system continuously learns from new data, enhancing its ability to detect and prevent fraud. This has resulted in a significant reduction in fraudulent transactions and improved security measures.

Automating Regulatory Compliance

The automation of regulatory compliance through AI has become a critical component in ensuring financial institutions adhere to ever-evolving regulations efficiently and accurately. By leveraging advanced data analysis and management capabilities, AI streamlines the complex processes of monitoring transactions and generating compliance reports. This automation reduces the manual workload traditionally associated with regulatory compliance while increasing accuracy and timeliness. For instance, IBM Watson's AI platform is used by financial institutions to automate these tasks, providing real-time alerts and generating precise compliance reports. This not only enhances adherence to regulations but also significantly mitigates the risk of compliance breaches, ensuring institutions remain compliant in real-time.

AI in Streamlining Regulatory Compliance

AI streamlines regulatory compliance by automating monitoring and reporting processes, ensuring adherence to financial regulations through advanced data analysis and management.

Automating Monitoring and Reporting

AI systems can automate the monitoring of transactions and the generation of compliance reports, reducing the manual workload and increasing accuracy. These systems ensure that financial institutions comply with regulatory requirements in real-time.

Example: AI-Driven Compliance Systems

Case Study: IBM Watson

IBM Watson's AI platform helps financial institutions automate regulatory compliance. By analyzing transaction data and monitoring for compliance breaches, Watson provides real-time alerts and generates accurate compliance reports. This has reduced the burden of manual compliance tasks and improved adherence to regulations.

Case Study: Ayasdi

Ayasdi uses AI to help financial institutions navigate complex regulatory requirements. The platform automates the analysis of large datasets, identifying compliance risks and generating reports that meet regulatory standards. This has streamlined compliance processes and reduced the risk of regulatory penalties.

Real-Time Data Processing and Predictive Analytics

AI's Capability for Real-Time Data Processing

AI can process vast amounts of data in real-time, providing up-to-the-minute financial insights and enabling more informed decision-making. This capability allows financial institutions to respond quickly to market changes and emerging opportunities.

Example: Real-Time Data Processing

Case Study: AlphaSense

AlphaSense uses AI to process and analyze real-time financial data, providing analysts with immediate insights into market trends and developments. The platform leverages LLMs to interpret complex financial information and generate actionable insights, helping analysts make faster and more informed decisions.

Predictive Analytics in Forecasting Market Trends

Predictive analytics enables financial institutions to forecast market trends, enhancing investment strategies and identifying emerging opportunities. By analyzing historical data and current market conditions, AI can predict future market movements with a high degree of accuracy.

Example: Predictive Analytics in Finance

Case Study: BlackRock's AI Lab

BlackRock's AI Lab uses predictive analytics to forecast market trends and inform investment strategies. The platform analyzes large datasets, including economic indicators and market sentiment, to generate accurate predictions. This has improved BlackRock's ability to identify investment opportunities and optimize portfolio performance.

Personalized Investment Advice

AI-Powered Personalized Investment Advice

AI-powered platforms offer personalized investment advice by analyzing individual financial goals, risk tolerance, and market conditions to tailor recommendations. These platforms provide investors with customized strategies that align with their unique financial objectives.

Example: AI-Powered Investment Platforms

Case Study: Personal Capital

Personal Capital uses AI to offer personalized investment advice. By analyzing user data, including financial goals and risk preferences, the platform provides tailored recommendations and automated portfolio management. This has helped investors achieve their financial objectives with greater confidence and efficiency.

Challenges and Ethical Considerations

Challenges and Ethical Considerations in AI Finance

The implementation of AI in finance comes with challenges and ethical considerations, including algorithmic transparency, data privacy, and the potential for systemic risk. Addressing these issues is critical to ensuring the responsible use of AI technologies.

Algorithmic Transparency

Ensuring transparency in AI algorithms is essential to building trust and accountability. Financial institutions must develop explainable AI models that provide clear insights into how decisions are made.

Example: Transparent AI Practices

Case Study: FICO

FICO uses explainable AI techniques to ensure transparency in its credit scoring systems. The platform provides detailed explanations of credit decisions, helping customers understand how their scores are determined and building trust in the process.

Data Privacy

Protecting customer data is a top priority for financial institutions using AI. Ensuring robust data privacy measures is essential to maintain client trust and comply with regulations.

Example: Data Privacy Measures

Case Study: GDPR Compliance

Financial institutions in Europe must comply with the General Data Protection Regulation (GDPR), which sets strict guidelines for data privacy. AI systems must be designed to protect user data, ensuring that personal information is handled securely and transparently.

Potential for Systemic Risk

AI systems can inadvertently introduce systemic risks if not properly managed. Ensuring that AI models are robust and resilient to market changes is critical to preventing unintended consequences.

Example: Managing Systemic Risk

Case Study: JPMorgan Chase

JPMorgan Chase uses AI to manage systemic risk by continuously monitoring market conditions and adjusting strategies in real-time. The platform's AI-driven risk management system helps identify potential threats and implement measures to mitigate risks, ensuring stability in volatile markets.

Measures to Mitigate Risks and Ensure Responsible AI Use

Financial institutions are taking measures to mitigate risks and ensure the responsible use of AI technologies. This includes developing ethical guidelines, implementing robust data privacy measures, and ensuring transparency in AI algorithms.

Example: Ethical AI Guidelines

Case Study: Partnership on AI

The Partnership on AI is a collaborative initiative that includes major financial institutions and technology companies. The partnership develops ethical guidelines for AI use, promoting transparency, fairness, and accountability in AI systems.

By leveraging these measures, financial institutions can harness the transformative potential of AI while addressing challenges and ethical considerations, ensuring that AI technologies are used responsibly and effectively in the financial sector.

Synergy Between AI and Blockchain Technology

The integration of AI and blockchain technology enhances data integrity, transparency, and security in financial transactions. Blockchain provides a secure and transparent ledger for recording transactions, while AI analyzes these transactions for potential fraud.

Enhancing Data Integrity and Transparency

Blockchain technology ensures that transaction data is immutable and transparent, providing a reliable source for AI analysis. This combination enhances trust and reduces the risk of data tampering.

Example: AI and Blockchain Integration

Case Study: IBM and Maersk

IBM and Maersk have developed a blockchain-based platform for global trade that uses AI to enhance security. The platform records transactions on a blockchain ledger, ensuring data integrity and transparency. AI analyzes the transaction data for anomalies and potential fraud, providing an additional layer of security.

Verifying Transactions with Blockchain and AI

Blockchain verifies the authenticity of transactions, while AI analyzes these transactions for suspicious activities. This integrated approach provides robust security for financial transactions.

Example: Blockchain and AI for Fraud Detection

Case Study: HSBC and ING

HSBC and ING have used blockchain and AI to streamline and secure trade finance transactions. The blockchain platform records transaction details, ensuring transparency and security. AI algorithms analyze the transaction data to detect potential fraud, enhancing the overall security of the trade finance process.

5.8. Challenges in AI-Driven Fraud Detection and Risk Management

Technical and Ethical Challenges

Implementing AI in fraud detection and risk management comes with several technical and ethical challenges:

- **Privacy Concerns:** The use of AI involves processing large amounts of personal and financial data, raising concerns about data privacy and security.

- **Data Bias:** AI models can inherit biases present in the training data, leading to unfair or discriminatory outcomes.

- **Regulatory Compliance:** Ensuring that AI systems comply with various regulatory requirements across different jurisdictions can be complex and challenging.

Strategies for Overcoming Challenges

To address these challenges, financial institutions can adopt several strategies:

- **Developing Transparent AI Models:** Ensuring transparency in AI models helps build trust and accountability. Explainable AI techniques can provide insights into how decisions are made.

- **Ensuring Data Protection:** Implementing robust data protection measures, such as encryption and access controls, can safeguard sensitive information and maintain privacy.

- **Mitigating Data Bias:** Using diverse and representative datasets for training AI models can help reduce bias. Regular audits and updates of AI systems can also ensure fairness and accuracy.

- **Adhering to Regulatory Standards:** Staying informed about regulatory requirements and working closely with compliance teams can help ensure that AI systems meet legal and ethical standards.

5.9. Developing a Retrieval-Augmented Generation (RAG) Application for Stock Recommendations Using LlamaIndex

Introduction

This guide demonstrates building a Retrieval-Augmented Generation (RAG) application for stock recommendations using LlamaIndex. By following this step-by-step approach, you can create a system that provides insightful stock recommendations based on financial documents.

Step-by-Step Guide to Building the Application

Step 1: Install Dependencies

CHAPTER 5   INNOVATIONS IN INVESTMENT BANKING

1. **Create and Activate a Virtual Environment:**

   ```
   cd stock_recommendation
   python -m venv .venv
   source .venv/bin/activate
   ```

2. **Explanation:** A virtual environment allows you to manage dependencies specific to this project without affecting other Python projects on your system. This ensures that the required packages are isolated and can be managed independently.

3. **Install Required Packages:**

   ```
   pip install llama-index flask flask-cors transformers torch chromadb psycopg2
   ```

   1. **Explanation:**

      - llama-index: Provides the core functionalities for creating and querying indexes.

      - flask and flask-cors: Flask is a lightweight WSGI web application framework, and Flask-CORS allows handling Cross-Origin Resource Sharing (CORS), making the API accessible from different domains.

      - transformers and torch: These libraries are crucial for working with LLMs (Large Language Models). Transformers provide tools and pre-trained models, while Torch is a deep learning library.

      - chromadb: An embedding database for efficient storage and retrieval of vector embeddings.

      - psycopg2: A PostgreSQL database adapter for Python, used for interacting with PostgreSQL databases.

CHAPTER 5   INNOVATIONS IN INVESTMENT BANKING

Step 2: Run Ollama Llama2

1. **Download and Install Ollama:**

    ```
    # Download and install Ollama from the official repository
    # Verify installation
    ls /usr/local/bin/ollama
    ```

    - **Explanation:** Ollama is a framework that simplifies the deployment and management of LLMs on personal computers. This step ensures that the Ollama binary is correctly installed and available for use.

    - **Run the Llama2 model:**

        ```
        root@150bc5106246:/# ollama run llama2
        # After downloading and setup, reconnect to the LLM console
        root@c96f4fc1be6f:/# ollama run llama2
        ```

    Explanation: This step starts the Llama2 model using Ollama. Running the model ensures it is set up correctly and ready to process queries.

Test the Model:

```
root@c96f4fc1be6f:/# ollama run llama2
>>> what is stock recommendation
Stock recommendation involves analyzing market data, company performance, and financial indicators to provide advice on buying, holding, or selling stocks.
```

1. **Explanation:** Testing the model verifies that it is functioning correctly and can respond to queries, ensuring that the setup process has been successful.

Step 3: Set Up and Run the RAG Application

1. **Define Configurations in config.py:**

    ```
    import os

    INIT_INDEX = os.getenv('INIT_INDEX', 'false').lower() == 'true'
    INDEX_PERSIST_DIRECTORY = os.getenv('INDEX_PERSIST_DIRECTORY', "./data/chromadb")
    ```

CHAPTER 5   INNOVATIONS IN INVESTMENT BANKING

```
HTTP_PORT = os.getenv('HTTP_PORT', 7654)
MONGO_HOST = os.getenv('MONGO_HOST', 'localhost')
MONGO_PORT = os.getenv('MONGO_PORT', 27017)
MONGO_USER = os.getenv('MONGO_USER', 'testuser')
MONGO_PASS = os.getenv('MONGO_PASS', 'testpass')
```

- **Explanation:** Configuration settings are defined here to manage various parameters like database connections, index directories, and API port numbers. Using environment variables helps keep sensitive information secure and allows easy configuration changes without modifying the code.

- Implement **the HTTP API in api.py:**

```
from flask import Flask, jsonify, request
from flask_cors import CORS
import logging
import sys
from model import *
from config import *

app = Flask(__name__)
CORS(app)

logging.basicConfig(stream=sys.stdout, level=logging.INFO,
format='%(asctime)s - %(levelname)s - %(message)s')

@app.route('/api/question', methods=['POST'])
def post_question():
    json = request.get_json(silent=True)
    question = json['question']
    user_id = json['user_id']
    logging.info("post question `%s` for user `%s`", question,
    user_id)

    resp = chat(question, user_id)
    data = {'answer': resp}

    return jsonify(data), 200
```

```
if __name__ == '__main__':
    init_llm()
    index = init_index(Settings.embed_model)
    init_query_engine(index)

    app.run(host='0.0.0.0', port=HTTP_PORT, debug=True)
```

- **Explanation:** This script sets up the Flask application to handle HTTP requests. The /api/question endpoint accepts POST requests with a question and user ID, processes the question using the chat function, and returns the response. The init_llm, init_index, and init_query_engine functions are called to initialize the LLM, index, and query engine, respectively.

- Implement **the Model in model.py:**

```
import chromadb
import logging
import sys
from llama_index.llms.ollama import Ollama
from llama_index.embeddings.huggingface import HuggingFaceEmbedding
from llama_index.core import (Settings, VectorStoreIndex, SimpleDirectoryReader, PromptTemplate)
from llama_index.core import StorageContext
from llama_index.vector_stores.chroma import ChromaVectorStore

logging.basicConfig(stream=sys.stdout, level=logging.INFO, format='%(asctime)s - %(levelname)s - %(message)s')

global query_engine
query_engine = None

def init_llm():
    llm = Ollama(model="llama2", request_timeout=300.0)
    embed_model = HuggingFaceEmbedding(model_name="BAAI/bge-small-en-v1.5")

    Settings.llm = llm
    Settings.embed_model = embed_model
```

```python
def init_index(embed_model):
    reader = SimpleDirectoryReader(input_dir="./docs", recursive=True)
    documents = reader.load_data()

    logging.info("index creating with `%d` documents", len(documents))

    chroma_client = chromadb.EphemeralClient()
    chroma_collection = chroma_client.create_collection("iollama")

    vector_store = ChromaVectorStore(chroma_collection=chroma_collection)
    storage_context = StorageContext.from_defaults(vector_store=vector_store)

    index = VectorStoreIndex.from_documents(documents, storage_context=storage_context, embed_model=embed_model)

    return index

def init_query_engine(index):
    global query_engine

    template = (
        "Imagine you are an advanced AI expert in financial analysis, with access to all current and relevant stock market data,"
        "analyst reports, and news articles. Your goal is to provide insightful, accurate, and concise stock recommendations based on this information.\n\n"
        "Here is some context related to the query:\n"
        "-----------------------------------------\n"
        "{context_str}\n"
        "-----------------------------------------\n"
        "Considering the above information, please respond to the following inquiry with detailed references to relevant data and analyses where appropriate:\n\n"
        "Question: {query_str}\n\n"
```

CHAPTER 5   INNOVATIONS IN INVESTMENT BANKING

```
            "Answer succinctly, starting with the phrase 'Based
            on current market analysis,' and ensure your response
            is understandable to someone without a financial
            background."
        )
        qa_template = PromptTemplate(template)

        query_engine = index.as_query_engine(text_qa_template=qa_
        template, similarity_top_k=3)

        return query_engine

    def chat(input_question, user):
        global query_engine

        response = query_engine.query(input_question)
        logging.info("got response from llm - %s", response)

        return response.response
```

- Explanation:
  - init_llm(): Initializes the LLM using Ollama and sets the embedding model from HuggingFace.
  - init_index(): Reads documents from a specified directory, creates an index using Chroma for vector storage, and logs the process.
  - init_query_engine(): Sets up a query engine with a custom prompt template that guides the LLM in providing relevant stock recommendations.
  - chat(): Processes user questions by querying the index and retrieving relevant information to generate a response.

- Run **the Application**:

```
# enable virtual environment in `ollama` source directory
>> cd stock_recommendation
>> source .venv/bin/activate
```

CHAPTER 5 INNOVATIONS IN INVESTMENT BANKING

```
# run application
2024-05-29 18:04:45,917 - INFO - index creating with `46` documents
2024-05-29 18:04:45,922 - INFO - Anonymized telemetry enabled. See https://docs.trychroma.com/telemetry for more information.
 * Serving Flask app 'api' (lazy loading)
 * Environment: production
   WARNING: This is a development server. Do not use it in a production deployment.
   Use a production WSGI server instead.
 * Debug mode: on
2024-05-29 18   04:47,151 - INFO - WARNING: This is a development server. Do not use it in a production deployment. Use a production WSGI server instead.
    * Running on all addresses (0.0.0.0)
    * Running on http://127.0.0.1:7654
    * Running on http://192.168.0.110:7654
   2024-05-29 18:04:47,151 - INFO - Press CTRL+C to quit
   2024-05-29 18:04:47,151 - INFO -  * Restarting with stat
   2024-05-29 18:04:52,083 - INFO - index creating with `46` documents
   2024-05-29 18:04:52,088 - INFO - Anonymized telemetry enabled. See https://docs.trychroma.com/telemetry for more information.
   2024-05-29 18:04:53,338 - WARNING -  * Debugger is active! huggingface/tokenizers: The current process just got forked, after parallelism has already been used. Disabling parallelism to avoid deadlocks...
   To disable this warning, you can either:
    - Avoid using `tokenizers` before the fork if possible
    - Explicitly set the environment variable TOKENIZERS_PARALLELISM=(true | false)
   2024-05-29 18:04:53,353 - INFO -  * Debugger PIN: 102-298-684
```

CHAPTER 5   INNOVATIONS IN INVESTMENT BANKING

Explanation:

This command starts the Flask application, initializing the index with 46 documents and setting up the necessary telemetry. The warning about the development server is a reminder that this setup is for testing and development purposes only and should be replaced with a production-grade WSGI server for deployment. The application is accessible via the specified local IP addresses.

Step 4: Post Question

1. Post a Question to the API:

```
# post question
>> curl -i -XPOST "http://localhost:7654/api/question" \
--header "Content-Type: application/json" \
--data '
{
  "question": "Which stocks should I buy?",
  "user_id": "user123"
}
'

# response from llm
HTTP/1.1 200 OK
{
  "answer": "Based on current market analysis, it is recommended to
  consider purchasing stocks in technology companies with strong
  earnings growth, such as Apple (AAPL) and Microsoft (MSFT).
  Additionally, look into renewable energy stocks like NextEra Energy
  (NEE) due to the growing focus on sustainable energy sources.
  It's also wise to diversify your portfolio by including healthcare
  stocks such as Johnson & Johnson (JNJ) to mitigate risks."
}
```

Explanation:

This command sends a POST request to the API endpoint with a JSON payload containing the question "Which stocks should I buy?" and a user ID. The server processes the question using the LLM, retrieves relevant data, and generates a response. The response includes stock recommendations based on current market analysis, demonstrating the system's ability to provide meaningful financial advice.

Further Steps:

1. Customizing and Scaling: You can customize the RAG application by integrating more sophisticated financial models and scaling it by deploying it on a cloud service with a production-ready WSGI server.

2. Enhancements: Adding features like user authentication, data visualization, and real-time stock market data integration can make the application more robust and user-friendly.

3. Advanced Features: Exploring advanced features like fine-tuning the LLM with domain-specific data can enhance the relevance and accuracy of the recommendations.

For further customization and advanced features, you can explore the full source code available on GitLab.

5.10. Summary

- Investment banking and trading sectors are undergoing significant transformation driven by AI, which enhances efficiency, accuracy, and personalization in service delivery.

- The integration of AI into these sectors addresses challenges like market volatility, regulatory pressures, and the need for continuous innovation to meet evolving client demands.

- Large Language Models (LLMs) contribute to improved decision-making by analyzing vast datasets, enabling the identification of trends, market movements, and personalized financial strategies.

- AI-driven tools in investment banking automate routine tasks, allowing financial professionals to focus on strategic activities while ensuring more precise risk management.

- Portfolio management has been revolutionized by AI, which now allows for the creation of tailored investment strategies and dynamic asset allocation based on real-time data.

- AI's role in customer service is enhanced through AI-powered chatbots and virtual assistants, providing instant, personalized support and improving overall customer satisfaction.

- AI significantly improves fraud detection and risk mitigation by analyzing transaction patterns, identifying anomalies, and enabling real-time risk assessment, thus enhancing financial stability.

- Regulatory compliance in the financial sector is increasingly managed through AI, which automates monitoring and reporting, reducing the manual workload and increasing accuracy in real-time compliance.

- AI is driving innovations in financial products, offering personalized solutions that align closely with individual client needs and evolving market conditions.

- The future of financial services is increasingly reliant on AI, necessitating a balanced approach that leverages AI's strengths while addressing ethical considerations, transparency, and the need for continuous adaptation by professionals.

CHAPTER 6

Transformative Practices in Modern Financial Services

6.1. Introduction

Financial services are undergoing a transformative shift, driven by advancements in artificial intelligence (AI). This chapter delves into the significant changes AI brings to financial advisory services, enhancing efficiency, accessibility, and personalization. The chapter is structured to provide a comprehensive overview of both traditional financial advisory methods and the innovative AI-driven solutions reshaping the industry. Traditional financial advisory services have long relied on personalized advice through manual analysis. Financial advisors engage directly with clients to understand their financial goals and circumstances, a process that is both time-consuming and resource-intensive. Consequently, these services come with high costs and limited accessibility, often catering primarily to high-net-worth individuals. Additionally, traditional methods face challenges in scalability and consistency, as the quality of advice can vary significantly based on the advisor's expertise and availability.

AI technologies, including machine learning, natural language processing (NLP), and predictive analytics, are revolutionizing the financial advisory landscape. By automating complex data analysis and decision-making processes, AI significantly boosts efficiency and accuracy in financial planning. These technologies enable more personalized and scalable financial guidance, democratizing access to high-quality financial advice and making it available to a broader audience. This transformation sets the stage for a new era in financial planning, where advanced algorithms and data-driven insights play a central role.

Robo-advisors are automated platforms that use algorithms to provide financial advice based on clients' financial goals, risk tolerance, and investment preferences. These platforms collect data through online questionnaires and employ sophisticated algorithms to recommend tailored investment strategies. Robo-advisors offer benefits such as lower fees, 24/7 availability, and the ability to serve a broader client base, including those underserved by traditional advisors. This chapter explores these themes in detail, providing insights into the evolving financial advisory services landscape and the pivotal role of AI in shaping its future.

6.2. Overview of the Traditional Financial Advisory Landscape

Traditional financial advisory services have long been characterized by personalized financial advice provided through manual analysis. This process involves financial advisors working directly with clients to understand their financial goals and circumstances, which is both time-consuming and resource-intensive. As a result, these services come with high costs and limited accessibility, often catering primarily to high-net-worth individuals. Furthermore, traditional advisory methods face challenges in scalability and consistency, as the quality of advice can vary significantly depending on the advisor's expertise and availability.

Introduction to the Transformative Role of AI

AI technologies, including machine learning, natural language processing (NLP), and predictive analytics, are transforming the financial advisory landscape. By automating complex data analysis and decision-making processes, AI significantly increases efficiency and accuracy in financial planning. These technologies enable more personalized and scalable financial guidance, democratizing access to high-quality financial advice and making it available to a broader audience. This shift sets the stage for a new era in financial planning and advisory, where advanced algorithms and data-driven insights play a central role.

The Rise of Robo-advisors

Robo-advisors are automated platforms that use algorithms to provide financial advice based on clients' financial goals, risk tolerance, and investment preferences. These platforms collect data through online questionnaires and employ sophisticated algorithms to recommend investment strategies tailored to individual needs. Robo-advisors continuously monitor and rebalance portfolios, ensuring that investment strategies remain aligned with clients' objectives.

Benefits of Robo-advisors

Robo-advisors offer several benefits over traditional financial advisory services. They typically charge lower fees, making financial advice more affordable and accessible to a wider audience. The convenience of 24/7 availability and easy-to-use online platforms further enhances their appeal. Additionally, robo-advisors can serve a broader client base, including those who may have been underserved by traditional advisors, by providing entry-level financial advice and personalized investment strategies.

Framework for building robo driver using LLM:

1. **Define Business Objectives and Compliance Requirements**

    - **Target Market Identification**: Use LLMs to analyze market research data, consumer feedback, and financial trends. This helps to refine the target demographic and tailor the platform's offerings to specific customer segments.

    - **Regulatory Compliance**: Integrate LLMs to assist in understanding and implementing local and international financial regulations. Use LLMs to automate the extraction of relevant regulatory information, which can be embedded into the platform's compliance workflows.

2. **Design the System Architecture**

    - **Backend Infrastructure**: Deploy your backend on cloud platforms like AWS or Google Cloud. LLMs can assist in designing scalable architectures by generating code snippets and recommending best practices for cloud resource management.

    - **Database Management**: Use LLMs to optimize database queries and structures, ensuring that data retrieval is efficient. Integrate databases like PostgreSQL or MongoDB to store client data securely, with LLMs providing natural language query interfaces for database interactions.

- **APIs for Data Integration**: LLMs can be used to write and refine API documentation and integration code, enabling seamless data flow between the robo-advisor and external financial data sources.

3. **Build the Core Algorithm**
   - **Risk Profiling**: Develop a risk profiling system where LLMs process user responses from an onboarding questionnaire, classifying them into risk categories. The model can be trained on historical user data to enhance profiling accuracy.
   - **Asset Allocation Strategy**: LLMs can help implement Modern Portfolio Theory (MPT) by generating code for portfolio optimization algorithms. They can also assist in continuously improving these algorithms through reinforcement learning.
   - **Portfolio Rebalancing**: Automate portfolio rebalancing using LLM-generated scripts that execute based on predefined thresholds or intervals. LLMs can provide explanations and alerts when rebalancing occurs, ensuring transparency.
   - **Tax Optimization**: Use LLMs to automate the detection of tax-loss harvesting opportunities, ensuring that the algorithm complies with local tax laws.

4. **Develop Client Interface**
   - **User Dashboard**: Leverage LLMs to create user-friendly interfaces. LLMs can generate UI/UX design suggestions and code snippets for a responsive dashboard that visualizes portfolio performance.
   - **Onboarding Process**: Implement an LLM-powered chatbot to guide users through the onboarding process. The chatbot can explain complex financial concepts and answer questions in real time.
   - **Client Communication**: Utilize LLMs to craft personalized communication strategies. LLMs can automate the generation of email content, push notifications, and chatbot interactions, providing users with updates and advice.

5. **Implement Advanced Analytics and AI**

    - **Machine Learning Models**: Use LLMs to develop and refine machine learning models for predicting market trends and optimizing asset allocations. LLMs can also generate explanations for these models, helping non-technical stakeholders understand the results.

    - **Natural Language Processing (NLP)**: Integrate LLMs for analyzing financial news, reports, and social media sentiment, feeding this data into the decision-making algorithms of your robo-advisor.

    - **Behavioral Analytics**: Implement LLMs to analyze user behavior and adjust risk profiles or recommendations accordingly, enhancing the personalization of the service.

6. **Testing and Validation**

    - **Backtesting**: Use LLMs to write backtesting scripts that simulate the performance of your investment strategies on historical data. Analyze the results and refine your algorithms based on the output.

    - **A/B Testing**: Automate A/B testing for different user interface designs or onboarding flows. LLMs can be used to analyze the results and suggest improvements.

    - **Compliance Testing**: Regularly use LLMs to parse regulatory updates and automate the testing of your platform's compliance with these regulations.

7. **Deployment**

    - **Continuous Integration/Continuous Deployment (CI/CD)**: Integrate LLMs into your CI/CD pipelines. Use them to automate code reviews, generate deployment scripts, and manage rollbacks if issues arise.

    - **Security Measures**: Use LLMs to identify potential security vulnerabilities in your code and suggest mitigations. Implement automated alerts for suspicious activity detected by the platform.

- **Performance Monitoring**: Set up monitoring tools like AWS CloudWatch, with LLMs assisting in analyzing logs and providing actionable insights for performance improvements.

8. **Post-deployment Support and Iteration**
    - **User Feedback Loop**: Use LLMs to analyze user feedback collected through surveys, chat interactions, and reviews. Implement suggested changes to improve user satisfaction.
    - **Regular Updates**: Schedule regular updates where LLMs assist in generating release notes, updating documentation, and automating the testing of new features.
    - **Compliance Audits**: LLMs can be used to generate compliance audit reports, ensuring that all legal and regulatory requirements are met continuously.

9. **Scale and Expand**
    - **Market Expansion**: Use LLMs to analyze potential new markets, identifying compliance requirements and user preferences. Generate localized content and interfaces based on the analysis.
    - **Partnerships**: Collaborate with fintech companies using LLMs to automate the due diligence process and streamline integration with third-party services.
    - **AI Enhancements**: Continuously train your AI models with new data. LLMs can help fine-tune models to improve accuracy, relevance, and the personalization of recommendations.

This framework leverages the capabilities of LLMs to streamline the development and deployment of a robo-advisor, enhancing both the technical and user experience aspects of the platform.

Personalized Financial Advice Through AI

AI enables more personalized financial advice by analyzing vast amounts of data to understand individual client needs and tailor recommendations accordingly. These technologies can process detailed financial histories, spending patterns, and investment behaviors to develop tailored investment strategies. AI also allows for real-time adjustments based on market conditions and significant life changes, ensuring that financial plans remain relevant and effective.

Examples of AI-Driven Tools and Platforms

AI-driven tools and platforms provide customized investment strategies, tax planning advice, and retirement planning. For instance, Betterment and Wealthfront use AI to manage investments and offer personalized financial planning. These platforms utilize algorithms to create tailored asset allocations, automate rebalancing, and optimize tax strategies. Similarly, TurboTax and H&R Block offer AI-assisted tax planning tools that provide personalized tax-saving recommendations. In retirement planning, platforms like Fidelity Go and Vanguard Digital Advisor use AI to develop customized retirement savings plans and adjust them as needed to meet long-term goals.

6.3. Impact on Long-Term Financial Planning

AI's Impact on Long-Term Financial Planning

AI significantly enhances long-term financial planning by enabling the simulation of various financial scenarios. AI-driven tools can model market fluctuations, economic changes, and personal life events to predict their impact on financial plans. This capability allows for proactive adjustments to investment strategies, ensuring that clients are better prepared for future uncertainties.

Predicting Future Financial Outcomes

AI's predictive analytics capabilities allow for accurate forecasting of future financial outcomes based on historical data and current market trends. Platforms like eMoney Advisor use AI to provide long-term projections for retirement and savings, helping clients understand the potential outcomes of different financial decisions. These insights enable both clients and advisors to make more informed decisions, enhancing the effectiveness of long-term wealth management strategies.

Enhancing Decision-Making in Wealth Management

AI-driven insights improve decision-making in wealth management by providing real-time data analysis and actionable recommendations. For example, Schwab Intelligent Portfolios Premium integrates AI to support advisory services with in-depth financial planning insights and continuous monitoring. This allows advisors to offer more precise and timely advice, improving client outcomes.

Enhancing Client-Advisor Interactions

AI tools and platforms augment the traditional client-advisor relationship by facilitating more informed discussions, continuous monitoring, and real-time updates. Advisors can use AI to analyze client data and provide personalized recommendations, enabling more meaningful and productive interactions. Real-time updates and continuous monitoring ensure that financial plans are always aligned with the latest market conditions and client needs, enhancing the overall advisory experience.

Empowering Advisors to Focus on Value-Added Services

AI empowers advisors to focus on value-added services and relationship building by automating routine analysis and data gathering. With AI handling the time-consuming tasks of data analysis and monitoring, advisors can dedicate more time to understanding their clients' unique needs, providing strategic advice, and building stronger relationships. This shift allows advisors to deliver higher quality and more personalized service, ultimately improving client satisfaction and loyalty.

Challenges and Ethical Considerations

Addressing Technical and Ethical Challenges

Integrating AI into financial planning and advisory services presents several technical and ethical challenges. Data privacy and security are critical concerns, as AI systems handle sensitive financial information. Ensuring compliance with data protection regulations, such as GDPR, is essential to protecting client data. Algorithmic bias is another significant issue, as biases in training data can lead to unfair or discriminatory outcomes. Additionally, there is a need for human oversight to ensure that AI-driven recommendations are accurate and appropriate.

Strategies for Navigating Challenges

To navigate these challenges, financial institutions must adopt transparent AI practices and maintain a human element in financial advisory. Developing explainable AI models that provide clear insights into decision-making processes helps build trust and accountability. Robust data protection measures, including encryption and access controls, are essential to safeguard sensitive information. Regular audits and updates to AI systems can help mitigate biases and ensure compliance with regulatory standards. Ensuring that human advisors review AI-generated recommendations can prevent errors and enhance the overall quality of advice.

Regulatory Landscape and Compliance

Overview of Regulatory Considerations

The use of AI in financial planning and advisory services must comply with various regulatory requirements. Financial institutions need to ensure that their AI systems adhere to data protection laws, such as GDPR and CCPA, which govern the handling and processing of personal information. Additionally, regulations from financial authorities, such as the SEC and FINRA in the United States, require that financial advice is suitable and in the best interest of clients.

Evolving Regulatory Framework

The regulatory framework for AI in financial services is evolving to address new challenges and ensure the responsible use of AI technologies. Financial institutions must stay informed about regulatory developments and ensure that their AI systems comply with legal standards. This includes implementing transparent and explainable AI practices, conducting regular compliance checks, and engaging with regulatory bodies to understand and address emerging requirements. Ensuring compliance not only protects clients but also enhances the credibility and trustworthiness of AI-driven financial advisory services.

6.4. The New Era of Financial Planning and Advisory

Traditional financial advisory services have long been characterized by personalized financial advice provided through manual analysis. This process involves financial advisors working directly with clients to understand their financial goals and circumstances, which is both time-consuming and resource-intensive. As a result, these services come with high costs and limited accessibility, often catering primarily to high-net-worth individuals. Furthermore, traditional advisory methods face challenges in scalability and consistency, as the quality of advice can vary significantly depending on the advisor's expertise and availability.

Introduction to the Transformative Role of AI

AI technologies, including machine learning, natural language processing (NLP), and predictive analytics, are transforming the financial advisory landscape. By automating complex data analysis and decision-making processes, AI significantly increases efficiency and accuracy in financial planning. These technologies enable more personalized and scalable financial guidance, democratizing access to high-quality financial advice and making it available to a broader audience. This shift sets the stage for a new era in financial planning and advisory, where advanced algorithms and data-driven insights play a central role.

Impact on Long-Term Financial Planning

AI significantly enhances long-term financial planning by enabling the simulation of various financial scenarios. AI-driven tools can model market fluctuations, economic changes, and personal life events to predict their impact on financial plans. This capability allows for proactive adjustments to investment strategies, ensuring that clients are better prepared for future uncertainties.

Predicting Future Financial Outcomes

AI's predictive analytics capabilities allow for accurate forecasting of future financial outcomes based on historical data and current market trends. Platforms like eMoney Advisor use AI to provide long-term projections for retirement and savings, helping clients understand the potential outcomes of different financial decisions. These insights enable both clients and advisors to make more informed decisions, enhancing the effectiveness of long-term wealth management strategies.

Enhancing Decision-Making in Wealth Management

AI-driven insights improve decision-making in wealth management by providing real-time data analysis and actionable recommendations. For example, Schwab Intelligent Portfolios Premium integrates AI to support advisory services with in-depth financial planning insights and continuous monitoring. This allows advisors to offer more precise and timely advice, improving client outcomes.

Enhancing Client-Advisor Interactions

AI tools and platforms augment the traditional client-advisor relationship by facilitating more informed discussions, continuous monitoring, and real-time updates. Advisors can use AI to analyze client data and provide personalized recommendations, enabling more meaningful and productive interactions. Real-time updates and continuous monitoring ensure that financial plans are always aligned with the latest market conditions and client needs, enhancing the overall advisory experience.

Empowering Advisors to Focus on Value-Added Services

AI empowers advisors to focus on value-added services and relationship building by automating routine analysis and data gathering. With AI handling the time-consuming tasks of data analysis and monitoring, advisors can dedicate more time to understanding their clients' unique needs, providing strategic advice, and building stronger relationships. This shift allows advisors to deliver higher quality and more personalized service, ultimately improving client satisfaction and loyalty.

Challenges and Ethical Considerations

Integrating AI into financial planning and advisory services presents several technical and ethical challenges. Data privacy and security are critical concerns, as AI systems handle sensitive financial information. Ensuring compliance with data protection regulations, such as GDPR, is essential to protect client data. Algorithmic bias is another significant issue, as biases in training data can lead to unfair or discriminatory outcomes. Additionally, there is a need for human oversight to ensure that AI-driven recommendations are accurate and appropriate.

Strategies for Navigating Challenges

To navigate these challenges, financial institutions must adopt transparent AI practices and maintain a human element in financial advisory. Developing explainable AI models that provide clear insights into decision-making processes helps build trust and accountability. Robust data protection measures, including encryption and access controls, are essential to safeguard sensitive information. Regular audits and updates to AI systems can help mitigate biases and ensure compliance with regulatory standards. Ensuring that human advisors review AI-generated recommendations can prevent errors and enhance the overall quality of advice.

Regulatory Landscape and Compliance

The use of AI in financial planning and advisory services must comply with various regulatory requirements. Financial institutions need to ensure that their AI systems adhere to data protection laws, such as GDPR and CCPA, which govern the handling and processing of personal information. Additionally, regulations from financial authorities, such as the SEC and FINRA in the United States, require that financial advice is suitable and in the best interest of clients.

Evolving Regulatory Framework

The regulatory framework for AI in financial services is evolving to address new challenges and ensure the responsible use of AI technologies. Financial institutions must stay informed about regulatory developments and ensure that their AI systems comply with legal standards. This includes implementing transparent and explainable AI practices, conducting regular compliance checks, and engaging with regulatory bodies to understand and address emerging requirements. Ensuring compliance not only protects clients but also enhances the credibility and trustworthiness of AI-driven financial advisory services.

This comprehensive overview of AI's transformative impact on financial planning and advisory services highlights the rise of robo-advisors, the personalization of financial advice, the impact on long-term financial planning and wealth management, and the enhancement of client-advisor interactions. It also addresses the challenges and ethical considerations in implementing AI and discusses the regulatory landscape and compliance issues related to the use of AI in financial services.

6.5. Overview of the Current State of AI in Financial Services

AI has rapidly become a cornerstone in the financial services industry, with technologies like machine learning, natural language processing (NLP), and blockchain integration driving significant advancements. Financial institutions are increasingly adopting AI to automate processes, enhance decision-making, improve customer experiences, and strengthen security. For instance, AI-powered chatbots and virtual assistants are now commonplace, providing 24/7 customer support and personalized financial advice.

Additionally, AI is revolutionizing areas such as fraud detection, risk management, and investment strategies. Predictive analytics enables financial institutions to foresee market trends and potential risks, allowing for more proactive management. Despite these advancements, the integration of AI in financial services is still in its nascent stages, and the full potential of these technologies is yet to be realized.

Setting the Stage for the Future

As AI continues to evolve, its role in financial services is expected to expand and deepen, fundamentally transforming the industry. The envisioned future of financial services involves AI not just enhancing but reimagining how financial institutions operate, interact with customers, and manage risks. AI-driven innovations will pave the way for more efficient, transparent, and personalized financial services, leading to a more inclusive and resilient financial ecosystem.

This future will be characterized by the seamless integration of AI across all aspects of financial services, from customer interactions and investment management to regulatory compliance and cybersecurity. As we look ahead, it is crucial to consider the ethical implications of AI, ensuring that these technologies are developed and deployed responsibly to avoid biases and protect consumer privacy.

CHAPTER 6   TRANSFORMATIVE PRACTICES IN MODERN FINANCIAL SERVICES

Ongoing Innovations in AI and Financial Services

Several cutting-edge AI innovations are currently transforming the financial industry:

- **Advanced Predictive Analytics**
  - AI-driven predictive analytics tools are becoming increasingly sophisticated, allowing financial institutions to analyze vast amounts of data and predict market trends, customer behavior, and potential risks with greater accuracy. These tools help in making more informed investment decisions, optimizing asset allocations, and managing risks proactively.

- **Example: Kensho Technologies**
  - Kensho, a subsidiary of S&P Global, uses AI to analyze financial markets and provide actionable insights. Their AI-powered platforms can predict market movements and identify investment opportunities, helping financial institutions make data-driven decisions.

- **Natural Language Processing (NLP) for Customer Interactions**
  - NLP technology is enhancing customer interactions by enabling more intuitive and responsive communication between customers and financial institutions. AI-powered chatbots and virtual assistants can handle customer inquiries, provide personalized financial advice, and assist with account management in real time.

- **Example: Bank of America's Erica**
  - Erica, Bank of America's AI-driven virtual assistant, uses NLP to assist customers with a range of banking tasks, from checking account balances to making payments and providing personalized financial tips.

- **Blockchain Integration for Security and Transparency**
  - The integration of AI with blockchain technology enhances the security and transparency of financial transactions. Blockchain provides a decentralized and immutable ledger, while AI can analyze transaction data for anomalies and potential fraud, ensuring a higher level of trust and security.

- **Example: IBM and Maersk's TradeLens**
    - TradeLens, a blockchain-based platform developed by IBM and Maersk, leverages AI to enhance global trade transparency and security. The platform uses blockchain to record transaction data and AI to analyze and verify the integrity of these transactions, reducing fraud and improving efficiency.

Pioneering Startups and Financial Institutions at the Forefront

Several startups and financial institutions are leading the charge in AI innovation within the financial sector:

- **Zest AI**
    - Zest AI uses machine learning to improve credit underwriting and reduce biases in lending. Their AI-driven models analyze a wider range of data points than traditional credit scoring methods, providing a more accurate assessment of credit risk and increasing access to credit for underserved populations.

- **Upstart**
    - Upstart leverages AI to offer more inclusive and efficient personal loans. By analyzing non-traditional data points such as education and employment history, Upstart's AI models provide a more comprehensive assessment of an applicant's creditworthiness, resulting in higher approval rates and lower default rates.

- **Ant Financial**
    - Ant Financial, a subsidiary of Alibaba, utilizes AI across its wide range of financial services, including payment processing, wealth management, and credit scoring. Their AI-powered risk management systems analyze transaction data in real time to detect and prevent fraud, ensuring the security and integrity of financial transactions.

6.6. Potential Disruptions Led by AI

Analysis of Potential Disruptions

AI is set to disrupt traditional financial models and institutions in several ways:

- **Shift Towards Decentralized Finance (DeFi)**
  - Decentralized finance leverages blockchain and smart contracts to offer financial services without traditional intermediaries like banks. AI can enhance DeFi by providing real-time risk assessments, automating contract execution, and improving security.

- **Example: Aave**
  - Aave is a DeFi platform that uses smart contracts to enable lending and borrowing without intermediaries. AI can be integrated to assess borrower risk in real time, providing more secure and efficient lending services.

- **Rise of Digital Currencies**
  - Digital currencies, including cryptocurrencies and Central Bank Digital Currencies (CBDCs), are transforming the way financial transactions are conducted. AI can enhance the security and efficiency of digital currency transactions, as well as provide real-time analytics for monitoring and compliance.

- **Example: Bitcoin**
  - AI can analyze blockchain transactions for patterns indicative of fraud or money laundering, helping to ensure the security and integrity of digital currency systems.

- **Transformation of Traditional Banking Roles**
  - AI is automating many tasks traditionally performed by bank employees, such as transaction processing, customer service, and compliance. This automation can free up human resources for more strategic roles focused on relationship management and advisory services.

- **Example: Wells Fargo**

  - Wells Fargo uses AI to automate routine tasks like transaction processing and fraud detection, allowing their employees to focus on more complex and high-value customer interactions.

Broader Economic and Societal Implications

The disruptions caused by AI in financial services will have broader economic and societal implications:

- **Economic Inclusion**

  - AI can provide financial services to underserved and unbanked populations through mobile technologies and microfinance solutions, promoting economic inclusion and reducing poverty.

- **Example: Tala**

  - Tala uses AI to provide microloans to individuals in emerging markets who lack access to traditional banking services. Their AI models assess creditworthiness based on mobile phone usage and other non-traditional data points.

- **Job Displacement and Creation**

  - While AI will automate many routine tasks, it will also create new job opportunities in areas such as AI development, data analysis, and compliance management. Financial institutions will need to invest in reskilling and upskilling their workforce to adapt to these changes.

- **Regulatory and Ethical Considerations**

  - The rise of AI in financial services will require updated regulatory frameworks to address issues such as data privacy, algorithmic bias, and the ethical use of AI. Financial institutions will need to work closely with regulators to ensure that AI technologies are used responsibly and transparently.

6.7. AI's Role in Fostering Sustainable and Inclusive Finance

AI can contribute to more sustainable financial practices by enabling ESG (Environmental, Social, and Governance) investing and green finance:

- **ESG Investing**
  - AI can analyze vast amounts of data to identify companies that meet ESG criteria, helping investors make more informed decisions and promote sustainable business practices.

- **Example: Truvalue Labs**
  - Truvalue Labs uses AI to analyze unstructured data from news articles, social media, and other sources to assess companies' ESG performance. This information helps investors identify sustainable investment opportunities.

- **Green Finance**
  - AI can support green finance initiatives by analyzing the environmental impact of projects and identifying investment opportunities that promote sustainability.

- **Example: ClimateAI**
  - ClimateAI uses machine learning to assess the climate risk of agricultural investments, helping financial institutions make more sustainable investment decisions.

Enhancing Financial Inclusion

AI has the potential to enhance financial inclusion by providing access to financial services for underserved or unbanked populations:

- **Mobile Technologies**
  - AI-driven mobile applications can provide financial services such as banking, lending, and insurance to individuals in remote or underserved areas.

- **Example: M-Pesa**
  - M-Pesa, a mobile money service in Kenya, uses AI to provide banking services to millions of unbanked individuals. AI-driven analytics help assess creditworthiness and manage risk, enabling more inclusive financial services.

- **Microfinance Solutions**
  - AI can improve the efficiency and accuracy of microfinance institutions, helping them provide small loans to entrepreneurs and small businesses in developing countries.

- **Example: Grameen Foundation**
  - The Grameen Foundation uses AI to optimize its microfinance operations, ensuring that loans are disbursed efficiently and that borrowers receive the support they need to succeed.

By leveraging AI to promote sustainable and inclusive finance, financial institutions can contribute to broader economic development and social well-being. However, it is crucial to ensure that these technologies are used responsibly and ethically to avoid unintended consequences and build trust with consumers and regulators alike.

6.8. Ethical Considerations and Responsible AI Development

Addressing Critical Ethical Issues

Ethical considerations are paramount in the development and deployment of AI-driven financial services. Key issues include

- **Privacy**
  - AI systems must ensure the confidentiality and security of sensitive financial and personal data. Financial institutions need to comply with data protection regulations and implement robust security measures to prevent data breaches.

- **Bias**
  - AI models can inadvertently perpetuate or exacerbate biases present in training data, leading to unfair or discriminatory outcomes. It is essential to identify and mitigate biases in AI algorithms to ensure fair and equitable treatment of all customers.

- **Accountability**
  - Transparent and explainable AI models are crucial for maintaining accountability in AI-driven decision-making. Financial institutions must ensure that AI systems provide clear and understandable explanations for their decisions, enabling auditability and regulatory compliance.

Strategies for Promoting Responsible AI Development

Promoting responsible AI development involves several strategies:

- **Ethical Guidelines**

  Establishing and adhering to ethical guidelines for AI development and deployment helps ensure that AI technologies are used responsibly. These guidelines should address issues such as data privacy, algorithmic fairness, and transparency.

- **Transparency in AI Algorithms**

  Developing explainable AI (XAI) models that provide clear and interpretable explanations of their decisions is critical. Techniques such as feature importance analysis and model interpretability can help enhance transparency.

- **Stakeholder Engagement**

  Engaging with stakeholders, including regulators, customers, and industry experts, helps ensure that AI technologies are developed and deployed in a manner that aligns with societal values and expectations. Regular feedback and collaboration can drive the adoption of best practices and standards.

Collaborative Efforts for a Resilient Financial Future

Need for Collaboration

CHAPTER 6   TRANSFORMATIVE PRACTICES IN MODERN FINANCIAL SERVICES

Ensuring the beneficial and responsible use of AI in finance requires collaboration among various stakeholders, including tech companies, financial institutions, regulatory bodies, and consumers. Collaborative efforts can help set standards, foster innovation, and address common challenges.

Examples of Successful Partnerships and Consortia

Several successful partnerships and consortia are driving innovation and setting standards in AI for financial services:

- **Partnership on AI**
  - The Partnership on AI is a collaborative effort involving tech companies, academia, and non-profit organizations aimed at promoting the responsible use of AI. The partnership focuses on areas such as fairness, transparency, and accountability in AI development.

- **Global Financial Innovation Network (GFIN)**
  - GFIN is a network of international regulators and financial institutions that collaborate to support financial innovation and improve regulatory frameworks. The network facilitates cross-border collaboration and helps develop common standards for AI in financial services.

6.9. Role of Prompt Engineering

The prompt plays (Table 6-1) a crucial role in preparing the analysis report by serving as a comprehensive guide for conducting a detailed and structured examination of a mutual fund. It ensures that all relevant aspects of the fund are thoroughly analyzed, enabling the financial planner to provide well-informed, evidence-based recommendations. Here's how the prompt facilitates the preparation of the analysis report:

1. **Setting the Context and Objectives:**

   The prompt begins by outlining the researcher's role and the scientific context. This helps establish the purpose of the analysis and clarifies the goals of the research. It ensures that the analysis is aligned with the needs of clients and focuses on providing actionable insights.

2. **Defining the Research Task:**

    By breaking down the research task into specific components (e.g., performance analysis, risk assessment, management evaluation), the prompt provides a clear roadmap for the research process. This structured approach helps the researcher systematically gather and analyze data, ensuring no critical aspect is overlooked.

3. **Providing a Real-World Example:**

    Using a real-world example, such as the Vanguard 500 Index Fund (VFIAX), the prompt grounds the analysis in a practical context. This makes the research more relatable and directly applicable to real-life financial planning scenarios, enhancing its relevance and usefulness.

4. **Structuring the Report:**

    The prompt outlines a detailed documentation format, including specific sections and subsections. This structure ensures the report is comprehensive and logically organized, making it easier for readers to follow and understand the analysis. Each section focuses on a distinct aspect of the fund, contributing to a holistic evaluation.

5. **Ensuring Professional Tone:**

    By specifying the use of a professional tone, the prompt guides the researcher to maintain clarity, precision, and objectivity in their writing. This enhances the credibility of the report and ensures it meets the standards expected in professional financial planning documents.

6. **Encouraging Data-Driven Analysis:**

    The prompt emphasizes the importance of supporting the analysis with data and empirical evidence. This focus on data-driven insights helps ensure the recommendations are grounded in factual information, increasing their reliability and validity.

CHAPTER 6   TRANSFORMATIVE PRACTICES IN MODERN FINANCIAL SERVICES

7. **Highlighting Key Areas of Focus:**

   By identifying critical areas such as risk assessment, management evaluation, and peer comparison, the prompt ensures that the analysis covers all essential dimensions of the mutual fund. This comprehensive approach helps identify both strengths and potential concerns, providing a balanced view.

8. **Facilitating Investor Suitability Assessment:**

   The prompt includes a section dedicated to evaluating the fund's suitability for different investor profiles. This helps tailor the analysis to the specific needs and goals of various clients, making the recommendations more personalized and actionable.

9. **Providing Clear Recommendations:**

   By summarizing findings and offering final recommendations, the prompt ensures that the report concludes with practical advice. This helps clients make informed decisions based on a thorough and well-reasoned analysis.

Example Analysis Report Using the Prompt

Title: Comprehensive Analysis of Vanguard 500 Index Fund (VFIAX)

Introduction: The Vanguard 500 Index Fund (VFIAX) is a cornerstone of many investment portfolios due to its broad market exposure and low-cost structure. This report aims to provide a detailed analysis of the fund, evaluating its performance, risk, management, and suitability for different investor profiles.

Fund Performance Analysis:

- **Historical Performance:** Over the past decade, VFIAX has consistently tracked the S&P 500 Index, delivering an average annual return of 10.5%.

- **Year-over-Year Returns:** The fund experienced positive returns in 8 out of the last 10 years, with the highest annual return of 32% in 2019.

- **Benchmark Comparison:** VFIAX's performance closely mirrors that of the S&P 500, with minimal tracking error.

Risk Assessment:

- **Volatility Measures:** The fund has a beta of 1.00, indicating it moves in line with the market. Its standard deviation over the past five years is 15%, reflecting moderate volatility.

- **Risk-Adjusted Returns:** The Sharpe ratio of 0.75 suggests that VFIAX offers reasonable returns for the level of risk taken.

- **Drawdown Analysis:** The maximum drawdown during the past decade was 35% during the 2008 financial crisis.

Management and Strategy Evaluation:

- **Fund Managers:** Managed by a team of experienced professionals at Vanguard, the fund benefits from a disciplined investment approach.

- **Investment Strategy:** The fund aims to replicate the performance of the S&P 500 Index by holding a diversified portfolio of large-cap U.S. stocks.

- **Portfolio Composition:** The fund's portfolio includes approximately 500 of the largest U.S. companies, with a turnover rate of around 3%.

Peer Comparison:

- **Similar Funds:** Compared to other S&P 500 index funds, VFIAX offers a competitive expense ratio of 0.04%.

- **Performance Metrics:** VFIAX's returns and risk measures are comparable to its peers, with no significant deviations.

Suitability for Investors:

- **Ideal Investor Profiles:** VFIAX is suitable for investors seeking long-term capital appreciation with moderate risk. It is particularly well-suited for retirement accounts and diversified portfolios.

- **Benefits and Drawbacks:** The fund's low expense ratio and broad market exposure are major benefits. However, its performance is tied to the U.S. stock market, which may not be ideal for investors seeking international diversification.

- **Alignment with Goals:** VFIAX aligns well with long-term financial goals, offering steady growth potential.

Conclusion: The Vanguard 500 Index Fund (VFIAX) is a robust investment option for those seeking broad market exposure with low costs. Its consistent performance, moderate risk profile, and experienced management make it a reliable choice for long-term investors. Based on this analysis, VFIAX is recommended for inclusion in diversified investment portfolios, particularly for retirement planning.

Example

Researcher Role/Persona

As a financial planner, your primary role is to assist clients in making informed decisions about their investments. Your responsibilities include analyzing various investment options, understanding market trends, and providing tailored advice to help clients achieve their financial goals.

Scientific Context

The scientific context involves a comprehensive analysis of a mutual fund. This includes examining the fund's performance, risk factors, management strategies, and alignment with clients' financial objectives. The aim is to provide evidence-based recommendations that maximize returns while minimizing risks.

Research Task

Your task is to conduct thorough research on a specific mutual fund. This involves

- Analyzing historical performance data
- Assessing the fund's risk profile
- Evaluating the management team's expertise and strategies
- Comparing the fund with benchmark indices and peer funds
- Identifying the fund's suitability for different investor profiles

Example

Example Mutual Fund: Vanguard 500 Index Fund (VFIAX)

Documentation Format

Title: Comprehensive Analysis of Vanguard 500 Index Fund (VFIAX)

Sections:

CHAPTER 6   TRANSFORMATIVE PRACTICES IN MODERN FINANCIAL SERVICES

1. **Introduction**
    - Importance of mutual funds in personal finance
    - Overview of Vanguard 500 Index Fund

2. **Fund Performance Analysis**
    - Historical performance metrics
    - Year-over-year returns
    - Comparison with benchmark indices

3. **Risk Assessment**
    - Volatility measures (e.g., standard deviation, beta)
    - Risk-adjusted returns (e.g., Sharpe ratio)
    - Drawdown analysis

4. **Management and Strategy Evaluation**
    - Background of fund managers
    - Investment strategy and philosophy
    - Portfolio composition and turnover

5. **Peer Comparison**
    - Analysis of similar funds in the same category
    - Comparative performance and risk metrics

6. **Suitability for Investors**
    - Investor profiles best suited for the fund
    - Potential benefits and drawbacks
    - Alignment with long-term financial goals

7. **Conclusion**
    - Summary of findings
    - Final recommendations

CHAPTER 6   TRANSFORMATIVE PRACTICES IN MODERN FINANCIAL SERVICES

Professional Tone

Maintain a professional and analytical tone throughout the document. Use clear, precise language and support your analysis with data and empirical evidence. Ensure that your writing is objective and free from biases, providing a balanced view of the mutual fund's strengths and weaknesses.

Prompt for Research:

Title: Comprehensive Analysis of Vanguard 500 Index Fund (VFIAX)

   Introduction: Begin with an engaging introduction that outlines the significance of mutual funds in personal finance and sets the stage for an in-depth examination of the Vanguard 500 Index Fund.

   Fund Performance Analysis: Analyze historical performance data and year-over-year returns, and compare the fund with benchmark indices to evaluate its performance over time.

   Risk Assessment: Assess the fund's risk profile by examining volatility measures such as standard deviation and beta, risk-adjusted returns like the Sharpe ratio, and perform a drawdown analysis.

   Management and Strategy Evaluation: Evaluate the expertise and strategies of the fund's management team, including their background, investment philosophy, portfolio composition, and turnover rates.

   Peer Comparison: Conduct a comparative analysis of the Vanguard 500 Index Fund with similar funds in the same category, focusing on performance and risk metrics.

   Suitability for Investors: Identify the investor profiles that would benefit most from investing in the Vanguard 500 Index Fund. Discuss potential benefits and drawbacks and how the fund aligns with various long-term financial goals.

   Conclusion: Summarize your findings and provide final recommendations based on your comprehensive analysis.

   Table 6-1 shows the various financial planner personas and their related prompts.

Table 6-1. Prompt Engineering

Financial Planner Role	Client Focus	Prompt for Preparing Advice
Wealth Manager	High-net-worth individuals	"Conduct a comprehensive review of the client's portfolio, focusing on asset diversification, tax optimization strategies, and estate planning. Provide recommendations for preserving and growing their wealth while minimizing risks."
Retirement Specialist	Pre-retirees and retirees	"Evaluate the client's current retirement savings and projected expenses. Develop a plan that includes sustainable withdrawal strategies, Social Security optimization, and healthcare cost management to ensure a comfortable retirement."
Investment Advisor	Individual investors	"Analyze the client's investment goals and risk tolerance. Create a diversified portfolio tailored to their needs, including recommendations for specific stocks, bonds, and mutual funds. Monitor and adjust the portfolio as needed."
Financial Coach	Individuals needing financial literacy and habit building	"Assess the client's current financial situation, including income, expenses, and debt. Provide guidance on budgeting, saving, and debt management. Help the client set and achieve short-term and long-term financial goals."
Tax Planner	Clients needing tax advice	"Review the client's current tax situation and identify opportunities for tax savings. Provide strategies for tax-efficient investments, retirement contributions, and deductions. Ensure compliance with all relevant tax laws and regulations."
Estate Planner	Clients focused on legacy planning	"Assess the client's estate planning needs, including wills, trusts, and beneficiary designations. Develop a comprehensive estate plan that minimizes tax liabilities and ensures the smooth transfer of assets to heirs and beneficiaries."

(continued)

CHAPTER 6 TRANSFORMATIVE PRACTICES IN MODERN FINANCIAL SERVICES

Table 6-1. (*continued*)

Financial Planner Role	Client Focus	Prompt for Preparing Advice
Insurance Specialist	Clients needing risk management	"Evaluate the client's insurance needs, including life, health, and property insurance. Recommend appropriate coverage levels and policies to protect against potential risks and ensure financial security for the client and their family."
Corporate Financial Planner	Business owners and corporate clients	"Analyze the financial health of the client's business. Provide strategic advice on financial planning, including employee benefits, retirement plans, risk management, and succession planning. Help optimize business operations and growth."
Ethical Financial Planner	Clients seeking ethical investments	"Understand the client's ethical values and investment preferences. Research and recommend socially responsible investment opportunities that align with their values, focusing on ESG criteria and impact investing."
Technology-Driven Planner	Tech-savvy clients	"Leverage fintech tools to analyze the client's financial situation. Provide automated investment solutions and digital financial planning advice. Stay updated with the latest fintech trends to offer innovative and efficient services."

One-Shot Prompt

Prompt:

"As a financial planner, analyze the retirement readiness of a client aged 50 with current retirement savings of $300,000, an annual income of $100,000, and a desired retirement age of 65. Consider their current savings rate, expected Social Security benefits, investment growth rate, and projected expenses in retirement. Provide a detailed plan to ensure they achieve their retirement goals."

Role of the Prompt:

The one-shot prompt provides a detailed and specific scenario for the financial planner to address in a single attempt. It includes key client information and specific factors to consider, ensuring a focused and comprehensive analysis. This approach helps in crafting a precise and actionable retirement plan without needing multiple iterations or examples.

Few-Shot Prompt

Prompt:

"Analyze the financial planning scenarios below and provide suitable advice for each:

1. A 35-year-old professional earning $80,000 annually, with $20,000 in savings, aiming to buy a house in 5 years.

2. A 45-year-old entrepreneur with a fluctuating income, seeking advice on stabilizing finances and planning for their child's education.

3. A 60-year-old nearing retirement with $500,000 in savings and moderate risk tolerance, needing a plan for sustainable withdrawals.

For each scenario, include considerations such as savings strategies, investment recommendations, risk management, and specific financial goals."

Role of the Prompt:

The few-shot prompt provides multiple scenarios with varying client profiles and financial goals. This allows the financial planner to demonstrate versatility and adaptability in offering tailored advice. By addressing several examples, the planner can showcase a range of skills and strategies applicable to different client needs, ensuring a broad and well-rounded approach to financial planning.

RCI Technique Prompt (Role, Context, Instruction)

Prompt:

Role: Financial Planner

Context: A client, 40 years old, earning $90,000 annually, has $50,000 in savings, and is concerned about both short-term financial stability and long-term wealth accumulation. They have moderate risk tolerance and are interested in investing but lack detailed knowledge.

Instruction: Develop a comprehensive financial plan that includes budgeting advice, short-term and long-term investment strategies, risk management, and educational resources to improve the client's financial literacy. Ensure the plan is actionable and tailored to their income, savings, and risk tolerance.

Role of the Prompt:

The RCI technique prompt clearly defines the financial planner's role, provides a detailed context of the client's financial situation, and gives specific instructions on what the planner should address. This structured approach ensures clarity and focus, guiding the planner to create a detailed, personalized financial plan that meets the client's needs. The RCI technique helps in breaking down the task into manageable parts, ensuring that all aspects of the client's financial situation are considered and addressed comprehensively.

Summary of Each Prompt's Role

1. **One-Shot Prompt:**
   - Provides a detailed, singular scenario for in-depth analysis.
   - Ensures focused and comprehensive advice in one attempt.
   - Ideal for precise and actionable planning without needing multiple examples.

2. **Few-Shot Prompt:**
   - Offers multiple scenarios to demonstrate versatility and adaptability.
   - Allows for showcasing a range of skills and strategies.
   - Ensures a broad and well-rounded approach to financial planning.

3. **RCI Technique Prompt:**
   - Clearly defines the role, context, and instructions.
   - Ensures a structured and focused approach to addressing all aspects of the client's financial situation.
   - Guides the planner to create a detailed and personalized plan, considering specific client needs.

These prompts help financial planners provide targeted, effective, and comprehensive advice tailored to different client scenarios and needs.

Conclusion

The financial services industry is on the brink of a significant transformation driven by artificial intelligence (AI). Traditional financial advisory models, characterized by manual analysis and high costs, are giving way to more efficient, accessible, and personalized AI-driven solutions. AI technologies such as machine learning, natural language processing (NLP), and predictive analytics are revolutionizing the way financial advice is delivered, making it possible to provide high-quality, scalable financial guidance to a broader audience.

Robo-advisors exemplify this shift, offering automated, algorithm-based financial advice that is tailored to individual client needs. These platforms not only reduce costs and enhance accessibility but also ensure continuous portfolio management and real-time adjustments. AI-driven tools further enable personalized financial advice by analyzing vast amounts of data, allowing for more accurate and relevant recommendations that adapt to changing market conditions and personal circumstances.

While the benefits of AI in financial services are substantial, the integration of these technologies also brings challenges, particularly in terms of data privacy, security, and ethical considerations. Ensuring compliance with data protection regulations, mitigating algorithmic biases, and maintaining a human element in financial advisory are crucial for building trust and accountability.

The regulatory landscape for AI in financial services is evolving, and financial institutions must stay informed and compliant with emerging standards. By adopting transparent and explainable AI practices, financial advisors can enhance the credibility and trustworthiness of their services.

In conclusion, AI is set to redefine financial advisory services, offering unprecedented opportunities for efficiency, personalization, and inclusivity. As AI technologies continue to advance, they will play an increasingly central role in financial planning and wealth management, paving the way for a more innovative and accessible financial ecosystem.

Researcher Role/Persona As a financial planner, your primary role is to assist clients in making informed decisions about their investments. Your responsibilities include analyzing various investment options, understanding market trends, and providing tailored advice to help clients achieve their financial goals.

Scientific Context The scientific context involves a comprehensive analysis of a mutual fund. This includes examining the fund's performance, risk factors, management strategies, and alignment with clients' financial objectives. The aim is to provide evidence-based recommendations that maximize returns while minimizing risks.

Research Task Your task is to conduct thorough research on a specific mutual fund. This involves

- Analyzing historical performance data
- Assessing the fund's risk profile
- Evaluating the management team's expertise and strategies
- Comparing the fund with benchmark indices and peer funds
- Identifying the fund's suitability for different investor profiles

Example

Example Mutual Fund: Vanguard 500 Index Fund (VFIAX)

Documentation Format

Title: Comprehensive Analysis of Vanguard 500 Index Fund (VFIAX)

Sections:

- **Introduction**
  - Importance of mutual funds in personal finance
  - Overview of Vanguard 500 Index Fund
- **Fund Performance Analysis**
  - Historical performance metrics
  - Year-over-year returns
  - Comparison with benchmark indices
- **Risk Assessment**
  - Volatility measures (e.g., standard deviation, beta)
  - Risk-adjusted returns (e.g., Sharpe ratio)
  - Drawdown analysis

Management and Strategy Evaluation

Background of fund managers

Investment strategy and philosophy

Portfolio composition and turnover

Peer Comparison

Analysis of similar funds in the same category

Comparative performance and risk metrics

Suitability for Investors

Investor profiles best suited for the fund

Potential benefits and drawbacks

Alignment with long-term financial goals

Conclusion

Summary of findings

Final recommendations

6.10. Summary

- AI technologies, including machine learning, natural language processing (NLP), and predictive analytics, are revolutionizing the financial advisory landscape by automating complex data analysis and decision-making processes, thereby enhancing efficiency and personalization.

- Robo-advisors leverage algorithms to provide financial advice based on clients' financial goals, risk tolerance, and investment preferences, offering benefits such as lower fees, 24/7 availability, and the ability to serve a broader client base.

CHAPTER 6   TRANSFORMATIVE PRACTICES IN MODERN FINANCIAL SERVICES

- AI-driven tools enable more personalized financial advice by analyzing vast amounts of data to understand individual client needs, leading to tailored investment strategies and real-time adjustments based on market conditions.

- The rise of robo-advisors reflects a shift towards more scalable and consistent financial advice, making high-quality advisory services more accessible to a wider audience, including those previously underserved by traditional advisors.

- AI enhances long-term financial planning by enabling the simulation of various financial scenarios, allowing for proactive adjustments to investment strategies to better prepare clients for future uncertainties.

- AI's predictive analytics capabilities allow for accurate forecasting of future financial outcomes, helping both clients and advisors make more informed decisions and improve the effectiveness of wealth management strategies.

- AI tools augment traditional client-advisor relationships by facilitating more informed discussions, continuous monitoring, and real-time updates, leading to more meaningful and productive interactions.

- Integrating AI into financial planning and advisory services presents challenges such as data privacy, security, and algorithmic bias, requiring robust measures to ensure compliance with regulations and maintain the integrity of AI-driven recommendations.

- The regulatory framework for AI in financial services is evolving, necessitating that financial institutions stay informed about developments and implement transparent and explainable AI practices to ensure compliance and build trust.

- Collaborative efforts among financial institutions, tech companies, and regulatory bodies are essential for setting standards, fostering innovation, and ensuring the responsible and ethical use of AI in financial services.

CHAPTER 7

The Evolution of Insurance in the Digital Age

7.1. Introduction to Insurance Product Innovation

The traditional insurance product landscape has been largely static, relying on standardized policies designed to cover a broad spectrum of risks. However, the rapidly evolving risk environment and changing consumer expectations necessitate a shift towards more innovative insurance products. Today's consumers demand flexibility, personalization, and coverage for emerging risks that traditional policies do not adequately address. This section explores the emergence of new insurance products and services that cater to specific modern-day needs and preferences, leveraging cutting-edge technologies such as Generative AI and Large Language Models (LLMs) to enhance innovation

The Digital Transformation of the Insurance Industry

Digital transformation has emerged as a driving force reshaping industries worldwide, significantly altering traditional business models and operational efficiencies. The insurance sector, traditionally reliant on historical data, manual processes, and one-size-fits-all policies, is now undergoing a profound shift due to technological advancements, including generative AI. This section (Table 7-1) explores how digital innovation, particularly generative AI, is revolutionizing the insurance industry, highlighting the opportunities it presents, the challenges it poses, and its overall impact on industry practices.

Chapter 7 The Evolution of Insurance in the Digital Age

Historical Context

The insurance industry has long been rooted in conventional practices, relying heavily on historical data and actuarial tables to assess risk and determine policy premiums. Manual processes dominated underwriting, claims processing, and customer interactions, often resulting in inefficiencies and delayed responses. Policies were largely standardized, offering little customization to meet individual needs. Generative AI is now transforming these traditional approaches. For example, generative AI models can create synthetic data sets to simulate various risk scenarios, enhancing the accuracy of risk assessments even when historical data is scarce or outdated.

Catalysts for Digital Transformation

Several factors are driving the digital transformation of the insurance industry, with generative AI playing a crucial role. Technological advancements such as artificial intelligence (AI), the Internet of Things (IoT), and blockchain are at the forefront. Generative AI, specifically, enables insurers to analyze vast amounts of data quickly and accurately, improving risk assessment and underwriting processes. For instance, AI can generate detailed risk profiles based on real-time data from IoT devices like smart home sensors and wearable health monitors. Additionally, blockchain offers secure, transparent record-keeping, enhancing trust and reducing fraud. Changing consumer behaviors also drive this transformation, as modern consumers expect seamless, personalized experiences facilitated by generative AI's ability to tailor interactions and recommendations dynamically.

Digital Transformation: Scope and Significance

Digital transformation in the insurance sector encompasses a wide range of innovations, including generative AI applications from data analytics and personalized policies to automated claims processing and fraud detection. Insurers leverage big data analytics to gain deeper insights into customer behaviors and preferences, enabling them to offer more tailored and relevant products. Generative AI can create personalized insurance policies based on individual risk profiles, moving away from the traditional one-size-fits-all approach. For example, AI-generated personalized health insurance plans can adjust coverage and premiums based on continuous health data from wearable devices.

Automation, powered by generative AI, is streamlining claims processing, reducing the time and effort required to settle claims, and improving customer satisfaction. AI-driven chatbots and virtual assistants enhance customer engagement by providing instant support and information, often generating tailored responses based on the customer's history and queries. Additionally, predictive analytics and machine learning, supported by generative AI, detect fraudulent claims more effectively, saving costs and maintaining the integrity of the insurance system.

Opportunities Unleashed by Digital Innovation

Digital transformation, particularly through generative AI, presents numerous opportunities for the insurance industry. The creation of new insurance products, such as usage-based or on-demand insurance, allows insurers to tap into previously underserved markets. Generative AI can develop real-time risk assessment models, enabled by IoT devices and data analytics, allowing for dynamic pricing and proactive risk management. For instance, generative AI can simulate driving behaviors and conditions to offer personalized auto insurance premiums based on real-time data.

These innovations also enable insurers to better meet the needs of a diverse and evolving customer base. For example, telematics in auto insurance allows for personalized premiums based on driving behavior, appealing to safe drivers seeking lower rates. Health insurers can use generative AI to analyze wearable data and incentivize healthy behaviors, aligning with consumer trends towards wellness and preventive care. This AI can generate personalized wellness recommendations and insurance incentives, enhancing customer engagement and loyalty.

Challenges and Considerations

While digital transformation offers significant benefits, it also presents challenges, particularly with generative AI. Data privacy concerns are paramount, as insurers must handle vast amounts of sensitive personal information securely and ethically. Generative AI models require access to extensive data, raising concerns about data protection and privacy. Regulatory compliance is another critical consideration, as insurers must navigate complex and evolving regulations to ensure their digital initiatives meet legal requirements. The risk of technological obsolescence is also a concern. Insurers must continually invest in and update their digital infrastructure to stay competitive, which can be resource-intensive. Additionally, the integration of new technologies with legacy systems can be complex and disruptive.

Modern Approaches to Risk Analysis in the Digital Age

Traditional risk assessment methods in insurance have primarily relied on historical data and actuarial tables to predict potential risks and determine policy premiums. As the volume and variety of data have exponentially increased, these traditional methods have struggled to keep pace. Generative AI revolutionizes traditional risk assessment methodologies, leading to more sophisticated and dynamic pricing models that benefit both insurers and policyholders through greater accuracy and personalization. For instance, generative AI can simulate diverse risk scenarios based on current data trends, offering more precise risk evaluations.

CHAPTER 7 THE EVOLUTION OF INSURANCE IN THE DIGITAL AGE

The Role of Big Data in Risk Analysis

The exponential growth in data availability has significantly enhanced risk assessment models. Insurers now have access to a plethora of new data sources, such as social media, IoT devices, and other digital footprints, which contribute to creating a more detailed and accurate risk profile. Generative AI can process this influx of data, allowing for a more granular analysis of risk factors. For example, generative AI can create synthetic profiles from social media data to predict lifestyle choices and activities that impact risk levels, providing insurers with actionable insights.

Table 7-1. Approaches to Insurance

Aspect	Description	Examples
Historical Context	Traditional reliance on historical data and manual processes; generative AI enhances risk assessment with synthetic data.	Generative AI models create synthetic datasets for risk scenarios.
Catalysts for Digital Transformation	Driven by AI, IoT, blockchain; generative AI creates detailed risk profiles and personalized interactions.	AI-generated risk profiles from IoT data; personalized AI interactions.
Scope and Significance of Digital Transformation	Encompasses data analytics, personalized policies, automated claims; generative AI creates personalized policies and improves customer engagement.	AI-driven chatbots, predictive analytics; personalized health insurance plans.
Opportunities Unleashed by Digital Innovation	New insurance products, real-time risk assessment; generative AI models simulate scenarios for personalized premiums.	Simulated driving behaviors for auto insurance premiums.
Challenges and Considerations	Data privacy, regulatory compliance, technological obsolescence; generative AI requires extensive data and secure handling.	Extensive data needed for AI; secure data handling required.
Modern Approaches to Risk Analysis	Digital technologies revolutionize traditional methods; generative AI offers dynamic pricing models and accurate assessments.	AI models simulate diverse risk scenarios for precise evaluations.

(continued)

Table 7-1. (*continued*)

Aspect	Description	Examples
Role of Big Data in Risk Analysis	Access to new data sources like social media and IoT; generative AI processes data for granular risk analysis.	Generative AI uses social media data for risk insights.
Artificial Intelligence and Machine Learning	AI/ML analyzes vast datasets for patterns and trends; generative AI creates predictive models for risk and fraud detection.	Predictive models for claims and fraud detection.
Integration of IoT Devices	Real-time data collection from IoT devices; generative AI processes data for personalized, dynamic pricing models.	Telematics data analyzed for usage-based insurance.
Blockchain Technology in Risk Management	Enhances data integrity and security; generative AI ensures accurate, transparent records on blockchain.	Generative AI on blockchain for tamper-proof records.
Implications for Insurers	More accurate pricing models and improved segmentation; generative AI creates dynamic, real-time pricing.	Generative AI dynamic pricing adjusts in real time.
Implications for Policyholders	Personalized policies and recommendations; generative AI suggests practices to improve risk profiles.	AI suggests safe driving and healthy lifestyle changes.

Artificial Intelligence and Machine Learning Using Generative AI

AI and ML algorithms play a crucial role in analyzing vast datasets to identify patterns, trends, and anomalies that were previously undetectable. These technologies enable insurers to forecast potential risks and outcomes with greater precision. Generative AI, in particular, can create predictive models that continuously learn and adapt from new data, improving their accuracy over time. For instance, generative AI can forecast the likelihood of a claim being filed or detect fraudulent activities by generating models based on diverse datasets and evolving patterns.

CHAPTER 7   THE EVOLUTION OF INSURANCE IN THE DIGITAL AGE

The Integration of IoT Devices

IoT devices are revolutionizing risk assessment by providing real-time data collection. Generative AI can process this continuous monitoring data, allowing for more personalized and dynamic pricing models. In auto insurance, generative AI can analyze data from telematics devices to offer usage-based insurance policies that reflect the individual risk profile of each driver. Similarly, in health insurance, generative AI can analyze wearable data to monitor physical activity, heart rate, and other health metrics, allowing insurers to adjust premiums based on the policyholder's lifestyle and health habits. These personalized approaches benefit insurers by improving risk segmentation and encouraging policyholders to engage in risk-reducing behaviors:

1. **Real-Time Data Processing**: LLMs enable the analysis of real-time data collected from IoT devices, such as telematics in auto insurance or wearables in health insurance. By processing this continuous stream of data, LLMs can provide insights into behavioral patterns and risk factors, allowing insurers to adjust pricing models dynamically based on real-time information.

2. **Personalized Risk Profiles**: LLMs facilitate the creation of highly personalized risk profiles by analyzing diverse datasets from IoT devices. In auto insurance, for example, LLMs can interpret driving patterns from telematics data to assess individual driver behavior, offering usage-based insurance policies tailored to the specific risk profile of each driver.

3. **Dynamic Pricing Models**: Leveraging the predictive capabilities of LLMs, insurers can develop dynamic pricing models that adjust in real time according to the data received from IoT devices. This approach allows insurers to more accurately price premiums based on the actual risk posed by each policyholder, rather than relying solely on traditional risk assessment methods.

4. **Encouraging Risk-Reducing Behaviors**: LLMs can be used to provide personalized feedback to policyholders based on data from IoT devices, encouraging behaviors that reduce risk. For example, in health insurance, LLMs can analyze wearable data to provide insights and recommendations for healthier lifestyles, which can lead to lower premiums for policyholders who engage in health-positive behaviors.

5. **Improved Risk Segmentation:** The ability of LLMs to process and interpret large volumes of IoT-generated data enables more granular risk segmentation. Insurers can identify subgroups within their customer base that exhibit similar risk profiles, allowing for more targeted and effective risk management strategies, ultimately leading to better risk mitigation and financial outcomes for insurers.

Blockchain Technology in Risk Management

Blockchain technology offers significant potential for enhancing data integrity and security in risk analysis. Generative AI can work alongside blockchain to ensure the accuracy and reliability of risk-related data. For instance, generative AI can create tamper-proof records of transactions and claims on a blockchain ledger, facilitating transparent and secure data sharing among insurers. This integration improves collaboration and reduces duplication of efforts in risk assessment, enhancing overall efficiency.

Implications for Insurers

Modern approaches to risk analysis allow insurers to develop more accurate and dynamic pricing models. By leveraging big data, AI, IoT, and generative AI, insurers can refine their risk segmentation processes, leading to more precise underwriting and pricing strategies. For example, generative AI can create dynamic pricing models that adjust premiums in real time based on continuously updated risk data. This helps in reducing claims costs and enhances customer segmentation, allowing insurers to target and tailor their products more effectively. The operational benefits of these technologies include increased efficiency, reduced operational costs, and improved decision-making capabilities.

Implications for Policyholders

Policyholders stand to benefit significantly from these advancements. Personalized policies based on individual risk profiles can lead to potential premium reductions, particularly for those who engage in risk-reducing behaviors. Generative AI can offer personalized recommendations for policyholders to maintain or improve their risk profile, such as suggesting safe driving practices or healthy lifestyle changes. Enhanced customer service, through faster claims processing and more responsive interactions, also contributes to a better overall experience for policyholders. However, it is essential to address privacy and ethical considerations, as the collection and use of personal data must be handled with utmost care to protect policyholders' rights.

CHAPTER 7   THE EVOLUTION OF INSURANCE IN THE DIGITAL AGE

The digital transformation of the insurance industry marks a significant departure from traditional practices, offering a multitude of opportunities for innovation and improvement. By leveraging advanced technologies such as AI, big data, IoT, blockchain, and particularly generative AI, insurers can enhance operational efficiencies, provide more personalized services, and better manage risks. However, the journey towards digitalization also presents challenges, particularly in terms of data privacy, regulatory compliance, and the need for continuous technological investment. As the industry evolves, a balanced approach that embraces innovation while addressing these challenges will be key to unlocking the full potential of digital transformation in insurance.

7.2. Enhancing Customer Engagement and Personalized Policies in the Insurance Industry with Generative AI

Introduction to Customer-Centric Insurance Models
Traditionally, the insurance industry relied on standardized policies and limited interactions. Insurers often used face-to-face meetings, call centers, and paper-based processes, which were impersonal and inflexible. This approach led to customer dissatisfaction and low engagement. However, digital innovation, especially through generative AI, is shifting the industry towards customer-centric models, offering personalized experiences and policies.

Digital Platforms for Enhanced Customer Interaction
The evolution of customer interaction channels in the insurance industry has been profound. Traditional methods have given way to digital platforms like mobile apps and online portals, which offer dynamic and user-friendly interfaces. These platforms transform how insurers engage with customers, providing 24/7 accessibility and personalized experiences. For instance, policyholders can manage their policies at their convenience, update personal information, make premium payments, or file claims through these platforms. Generative AI enhances this by tailoring content and services based on user profiles and behaviors. For example, generative AI can create personalized notifications for a home insurance policyholder about weather-related risks and protection tips while generating driving tips based on individual driving patterns for auto insurance customers.

Leveraging Big Data for Personalization

Integrating big data analytics into insurance operations personalizes policies. Insurers now access vast data from social media, IoT devices, and transactions, allowing deep insights into customer behaviors and preferences. For example, telematics devices in vehicles collect data on driving behavior. Generative AI analyzes this data to create usage-based insurance policies with premiums based on actual risk. Similarly, health insurers use wearable data to monitor health metrics and offer tailored wellness programs. Generative AI can generate personalized health recommendations and insurance incentives based on this data, benefiting insurers through reduced claims and policyholders through improved health outcomes and potentially lower premiums.

LLM in Customer Engagement

AI and machine learning (ML) are transforming customer engagement in insurance. AI-driven chatbots and virtual assistants (Table 7-2) handle customer queries and support requests, providing instant responses and guiding users through policy management tasks. Leveraging natural language processing (NLP), these chatbots enhance customer experience with human-like interactions. Generative AI takes this further by generating personalized responses based on the customer's history and preferences. ML algorithms analyze customer data to predict future behaviors and needs, allowing insurers to proactively reach out with personalized offers and reminders, improving retention rates and fostering long-term relationships.

Table 7-2. Personalization of Insurance Products

Aspect	Description	Examples
Introduction to Customer-Centric Insurance Models	Insurance industry relied on standardized policies and limited interactions. Generative AI shifts industry towards personalized experiences and policies.	Generative AI shifts industry towards customer-centric models, offering personalized experiences and policies.
Digital Platforms for Enhanced Customer Interaction	Digital platforms like mobile apps and portals offer user-friendly interfaces. Generative AI tailors content based on user profiles.	Generative AI creates personalized notifications for home insurance about weather-related risks and protection tips.

(continued)

Table 7-2. (*continued*)

Aspect	Description	Examples
Leveraging Big Data for Personalization	Big data analytics personalizes policies. Generative AI analyzes data from social media and IoT for usage-based insurance policies.	Generative AI analyzes telematics data to create usage-based insurance policies with premiums based on actual risk.
AI and Machine Learning in Customer Engagement	AI and ML transform customer engagement. Generative AI generates personalized responses based on customer history and preferences.	Generative AI generates personalized responses based on customer history and preferences, enhancing customer experience.
Benefits of Personalized Policies for Policyholders	Personalized policies offer benefits for policyholders, ensuring affordable solutions. Generative AI explains how behavior impacts premiums and policy terms.	Generative AI explains how individual actions influence policy terms and costs, enhancing transparency and trust.
Challenges and Future Directions	Implementing AI and ML requires investment and handling data privacy concerns. Generative AI models require extensive data, raising concerns.	Generative AI models require extensive data, raising concerns about data protection and privacy.

Benefits of Personalized Policies for Policyholders

Personalized policies offer numerous benefits for policyholders, ensuring relevant and affordable insurance solutions. This approach leads to cost savings and higher satisfaction levels. Transparency and trust between insurers and policyholders also improve, as customers understand how their behavior impacts premiums and see the direct benefits of their actions. Generative AI can generate personalized explanations and insights into how individual actions influence policy terms and costs, further enhancing trust and engagement.

Challenges and Future Directions

Despite the advantages, implementing advanced data analytics, AI, and ML presents challenges. Significant investment in technology and talent is required, along with navigating data privacy concerns and regulatory compliance. Generative AI models require extensive data, raising concerns about data protection and privacy. Looking ahead, future customer-centric insurance will likely involve greater integration of digital technologies. As AI and ML evolve, insurers will offer more precise and proactive services, enhancing customer engagement and satisfaction. Blockchain technology could also provide new avenues for secure and transparent interactions, building even greater trust between insurers and policyholders. Generative AI can enhance these interactions by creating secure, transparent records on blockchain ledgers.

Leveraging Data Analytics and AI for Personalization
Insurers are increasingly utilizing data analytics and AI to gather comprehensive insights into individual customer behaviors, preferences, and risk profiles. By harnessing the power of these technologies, insurers can move beyond traditional demographic data, capturing real-time and behavioral data from various sources such as social media, IoT devices, transactional history, and digital interactions. For example, generative AI can analyze this wealth of data to create detailed customer profiles that predict future behaviors and needs.

Translating Data into Personalized Policies and Pricing
The insights gained from data analytics and AI are then used to develop personalized policies and pricing models. One prominent example is pay-as-you-drive insurance, where auto insurers like Progressive use telematics data to calculate premiums based on individual driving behaviors. Safe drivers who exhibit lower-risk behaviors, such as adhering to speed limits and avoiding sudden braking, can benefit from lower premiums. This not only rewards good driving habits but also provides a more equitable pricing model compared to traditional methods. Generative AI enhances this by generating tailored risk assessments and pricing strategies that reflect individual driver profiles accurately.

7.3. The Role of Technology in Customizing Policies

Machine Learning for Dynamic Pricing
Machine learning algorithms play a crucial role in enabling dynamic pricing models in insurance. These algorithms process vast amounts of data to predict future risks and adjust premiums in real time. For instance, in property insurance, machine learning models can analyze factors such as local crime rates, weather patterns, and property maintenance records to dynamically adjust coverage and pricing. Generative AI can further refine these models by simulating various risk scenarios and generating adaptive pricing models.

CHAPTER 7   THE EVOLUTION OF INSURANCE IN THE DIGITAL AGE

IoT for Real-Time Data Collection

The Internet of Things (IoT) is revolutionizing the way insurers collect and utilize data. IoT devices, such as smart home sensors, vehicle telematics, and wearable health monitors, provide real-time data that insurers can use to continuously assess risk and adjust policies. For example, a smart home equipped with sensors can detect potential hazards like water leaks or smoke, allowing the insurer to offer preventive maintenance tips or adjust coverage based on the real-time risk profile. In auto insurance, telematics devices monitor driving behavior, enabling insurers to offer pay-as-you-drive policies that reflect actual usage and driving habits. Generative AI can analyze this real-time data to create highly personalized and dynamic insurance solutions.

Blockchain for Secure Data Sharing

Blockchain technology offers a secure and transparent method for data sharing in the insurance industry. By using a decentralized ledger, blockchain ensures that all data transactions are immutable and tamper-proof, enhancing trust between insurers and policyholders. This technology is particularly useful in scenarios requiring data verification and fraud prevention. For example, in health insurance, blockchain can securely store and share medical records, ensuring that only authorized parties have access to sensitive information. Generative AI can enhance these processes by generating secure data sharing protocols that streamline verification and reduce fraud.

Case Study: Telematics in Auto Insurance

One notable example of leveraging technology for personalized policies is the use of telematics in auto insurance. Progressive's Snapshot program is a prime example. By installing a telematics device in their vehicles, policyholders allow Progressive to monitor their driving habits. The data collected, including speed, braking patterns, and time of day driven, is used to calculate a personalized premium. Drivers who demonstrate safe driving behaviors are rewarded with lower premiums, while those with riskier habits may see higher rates. Generative AI can enhance this by generating predictive models that adjust premiums in real time based on continuous driving data.

Case Study: Wearables in Health Insurance   Vitality, a health and life insurance provider, uses data from wearable fitness trackers to offer personalized wellness programs and incentives. Policyholders who engage in regular physical activity and meet specific health goals can earn rewards such as discounts on premiums, gym memberships, and even travel. By integrating wearable data into their health insurance model, Vitality encourages healthier lifestyles and reduces the overall risk of claims, creating a win-win situation for both the insurer and the insured. Generative AI can analyze wearable data to generate customized health recommendations and insurance incentives.

Case Study: Blockchain for Data Security   Guardtime has implemented a blockchain-based system for secure data management in the health insurance sector. By using blockchain to securely store and verify medical records, Guardtime ensures data integrity and reduces the risk of fraud. This technology allows insurers to quickly verify the authenticity of medical records during the claims process, speeding up settlements and reducing administrative costs. Policyholders benefit from enhanced data privacy and a more efficient claims experience. Generative AI can further optimize this system by generating secure, streamlined processes for data verification and fraud detection.

7.4. Improving Customer Loyalty and Satisfaction

Impact of Personalized Experiences and Policies
Personalized experiences and policies have a significant impact on customer loyalty and satisfaction. When customers receive services and products tailored to their specific needs and preferences, they are more likely to feel valued and understood by their insurer. This personalized approach fosters a stronger emotional connection between the customer and the insurer, leading to increased loyalty and higher satisfaction levels. Generative AI can generate highly personalized interactions and experiences, enhancing customer satisfaction and retention.

Evidence and Studies

Evidence from various studies supports the positive impact of personalization on customer satisfaction and loyalty. A survey by Salesforce found that 84% of customers say being treated like a person, not a number, is very important to winning their business. Furthermore, personalized recommendations can increase the likelihood of a purchase by 78%. In the insurance sector, personalized policies that reflect individual risk profiles and behaviors can lead to a more engaging and satisfying customer experience. Generative AI can create customized recommendations and insights that build trust and transparency, essential factors in maintaining long-term customer relationships.

7.5. Challenges in Personalization

Potential Challenges and Barriers

Implementing personalized insurance models comes with several challenges and barriers. One of the primary concerns is data privacy. As insurers collect and analyze vast amounts of personal data, they must ensure that this data is handled securely and in compliance with privacy regulations such as the General Data Protection Regulation (GDPR) and the California Consumer Privacy Act (CCPA). Failure to protect customer data can lead to significant legal and reputational risks. Generative AI models require extensive data, raising additional privacy concerns.

Regulatory hurdles also pose a challenge

The insurance industry is heavily regulated, and introducing new technologies and personalized models requires navigating a complex landscape of regulations and compliance requirements. Insurers must work closely with regulators to ensure that their personalized offerings meet all legal standards. Implementing the necessary technological infrastructure for personalization requires substantial investment. Insurers need to develop and maintain advanced data analytics platforms, integrate IoT devices, and secure blockchain systems. This can be resource-intensive and may require significant time and expertise.

Strategies for Overcoming Challenges

To overcome these challenges, insurers can employ several strategies. Firstly, they should prioritize customer privacy and trust by implementing robust data security measures and being transparent about data usage. This includes using encryption, secure data storage, and regular audits to ensure compliance with privacy regulations. Building strong relationships with regulators is also crucial. Insurers should engage in open dialogue with regulatory bodies to understand requirements and work collaboratively to develop compliant solutions. This proactive approach can help insurers stay ahead of regulatory changes and avoid potential legal issues. Investing in technology and talent is essential for successful personalization. Insurers should allocate resources to develop and maintain advanced data analytics platforms and hire skilled professionals who can manage and analyze data effectively. Partnering with technology providers or Insurtech startups can also provide access to cutting-edge solutions and expertise.

For instance, AXA, a leading global insurer, has taken significant steps to address the challenges of data privacy and regulatory compliance by implementing robust data security measures and maintaining transparency in their data usage practices. They employ encryption technologies and conduct regular audits to ensure compliance with regulations such as the General Data Protection Regulation (GDPR). This proactive stance has helped AXA build and maintain customer trust while staying ahead of regulatory requirements.

In terms of fostering strong regulatory relationships, Prudential Financial engages in continuous dialogue with regulatory bodies across various markets to understand evolving compliance requirements. By collaborating with regulators, Prudential ensures that their products and services meet all legal standards, reducing the risk of legal complications and positioning the company as a leader in regulatory compliance.

Moreover, companies like Allstate have heavily invested in technology and talent to drive personalization and innovation in their services. Allstate's deployment of advanced data analytics platforms allows them to process large volumes of data to offer personalized insurance products. They have also formed partnerships with Insurtech startups to access cutting-edge technologies and expertise, which has enabled them to innovate rapidly and maintain a competitive edge in the market.

By addressing these challenges and prioritizing customer privacy and trust, insurers can successfully implement personalized insurance models that enhance customer loyalty and satisfaction. This customer-centric approach not only benefits policyholders but also positions insurers for long-term success in a competitive and evolving industry.

Situation: Implementing Generative AI for Personalized Health Insurance Policies

CHAPTER 7   THE EVOLUTION OF INSURANCE IN THE DIGITAL AGE

Prompt 1: Question: How can we use generative AI to enhance personalized health insurance policies?

Chain of Thought: Generative AI can analyze vast amounts of health data collected from wearable devices to generate personalized health recommendations. By identifying patterns and predicting future health risks, the AI can tailor insurance policies to fit individual needs better. However, we must ensure data privacy and regulatory compliance.

Self-Critical Reflection: While the idea of using generative AI is promising, it's essential to address the potential risks of data breaches and ensure that the AI's predictions are accurate and unbiased. Additionally, the integration of AI with existing systems might pose technical challenges.

Prompt 2: Question: What types of data can we collect from wearable devices to improve health insurance policies?

Chain of Thought: Wearable devices can collect various data types, such as physical activity (steps, exercise duration), vital signs (heart rate, blood pressure), sleep patterns, and dietary habits. This data can help create a comprehensive health profile for each policyholder.

Self-Critical Reflection: Collecting diverse data types is beneficial, but we need to ensure that the data collected is accurate and reliable. Wearable devices may have limitations in data accuracy, which could affect the AI's recommendations. Regular calibration and validation of these devices are necessary.

Prompt 3: Question: How can generative AI provide personalized health recommendations based on wearable data?

Chain of Thought: Generative AI can analyze wearable data to identify patterns and trends. For example, it can detect irregular sleep patterns or insufficient physical activity and suggest improvements. It can also predict potential health risks, such as cardiovascular issues, and recommend preventive measures.

CHAPTER 7   THE EVOLUTION OF INSURANCE IN THE DIGITAL AGE

Self-Critical Reflection: While generative AI can provide valuable insights, it's crucial to ensure that the recommendations are practical and achievable for the policyholder. Unrealistic or overly complex suggestions might lead to poor adherence. A user-friendly approach is necessary for effective implementation.

> detect irregular sleep patterns or insufficient physical activity and suggest improvements. It can also predict potential health risks, such as cardiovascular issues, and recommend preventive measures.
> Self-Critical Reflection: While generative AI can provide valuable insights, it's crucial to ensure that the recommendations are practical and achievable for the policyholder. Unrealistic or overly complex suggestions might lead to poor adherence. A user-friendly approach is necessary for effective implementation.

> **Question**: How can generative AI provide personalized health recommendations based on wearable data?
>
> **Chain of Thought**: Generative AI can analyze data from wearables, such as heart rate, sleep patterns, and physical acti⌄ to identify patterns and trends that may not be immediately apparent. For example, it can detect irregular sleep cycles or

Prompt 4: Question: What are the potential benefits and challenges of using generative AI for personalized health insurance? **Chain of Thought:** Benefits include more accurate risk assessments, tailored policies, and incentives for healthier lifestyles. Challenges involve data privacy concerns, regulatory compliance, integration with legacy systems, and ensuring AI's fairness and transparency.

Self-Critical Reflection: While the benefits are significant, the challenges are equally daunting. Addressing data privacy and regulatory compliance requires robust measures and constant monitoring. Ensuring the AI's fairness and transparency is critical to maintaining trust among policyholders.

Prompt 5: Question: How can we address the ethical concerns related to using generative AI in health insurance? **Chain of Thought:** Ethical concerns include data privacy, consent, and potential bias in AI algorithms. To address these, we must implement strict data protection protocols, obtain explicit consent from policyholders, and regularly audit AI systems to identify and mitigate biases.

CHAPTER 7   THE EVOLUTION OF INSURANCE IN THE DIGITAL AGE

Self-Critical Reflection: Addressing ethical concerns is essential for the successful adoption of generative AI in health insurance. Continuous improvement and transparency in AI practices are necessary to build and maintain trust. Stakeholder engagement and feedback can also help address these concerns effectively.

7.6. Ethical Considerations and Pioneering Product Innovations in Insurance

Data Bias and Discrimination

The use of personal data for policy customization raises several ethical considerations. One primary concern is data bias, which can lead to discriminatory practices. AI and machine learning models can inadvertently perpetuate biases present in the data they are trained on, resulting in unfair treatment of certain groups. For instance, if an AI model is trained on historical data reflecting past discrimination, it may continue to disadvantage those same groups, whether based on race, gender, or socioeconomic status. Companies like Amazon have faced similar issues where their AI recruitment tool showed bias against female candidates because it was trained on resumes submitted over a decade, predominantly by men.

Ensuring Transparency in Data Use

Another ethical concern is ensuring transparency in how personal data is used. Policyholders need to be fully informed about what data is being collected, how it is being used, and who has access to it. For example, Generali, a major insurance company, has implemented transparent data policies where customers are regularly informed about data collection practices and usage. This approach helps build trust and ensures compliance with regulations such as the General Data Protection Regulation (GDPR).

Pioneering Product Innovations in Insurance

Introduction to Insurance Product Innovation

The traditional insurance product landscape has been largely static, relying on standardized policies designed to cover a broad spectrum of risks. However, the rapidly evolving risk environment and changing consumer expectations necessitate a shift towards more innovative insurance products. Today's consumers demand flexibility, personalization, and coverage for emerging risks that traditional policies do not adequately address. The objective of this section is to explore the emergence of new insurance products and services that cater to specific modern-day needs and preferences, leveraging cutting-edge technologies such as Generative AI and Large Language Models (LLMs) to enhance innovation.

Cyber Insurance: Safeguarding Digital Assets

With the increasing digitalization of business operations, cyber risks have become a significant concern. Data breaches, hacking, ransomware, and online fraud pose substantial threats to organizations, leading to financial losses, reputational damage, and operational disruptions. As these cyber threats proliferate, there is a growing demand for insurance products that offer protection against such risks.

Development of Cyber Insurance Policies

Cyber insurance policies have evolved to provide comprehensive coverage for a range of cyber-related incidents. These policies typically include coverage options such as data breach response, business interruption losses, cyber extortion, and liability coverage. The development and pricing of cyber insurance policies pose unique challenges. Unlike traditional risks, cyber risks are constantly evolving, making them difficult to predict and assess accurately. Insurers must consider factors such as business size, industry, cybersecurity measures, and data handled.

Leveraging Generative AI and LLMs

Generative AI and LLMs can significantly enhance the development and customization of cyber insurance products. By analyzing vast amounts of data on cyber incidents and their impacts, these technologies help insurers identify patterns and predict future risks more accurately. LLMs assist in drafting tailored policy documents and generating personalized recommendations for businesses based on their specific risk profiles. AI-driven simulations can model potential cyberattack scenarios, helping insurers refine coverage options and pricing strategies. For instance, AXA has utilized AI to better understand cyber risks and tailor their insurance products accordingly.

Market Growth and Acceptance

The market for cyber insurance has seen substantial growth, driven by increased awareness of cyber risks and the need for robust protection. Businesses of all sizes recognize the importance of cyber insurance as part of their risk management strategy. According to industry reports, the global cyber insurance market is projected to grow significantly, with more companies seeking coverage as regulatory requirements and cybersecurity threats intensify. Case studies highlight the growing acceptance and reliance on cyber insurance. For example, a mid-sized company that suffered a ransomware attack leveraged its cyber insurance policy to cover the ransom payment, restore its systems, and manage the associated legal and reputational fallout. This real-world example underscores the value of cyber insurance in mitigating the financial impact of cyber incidents.

On-Demand and Usage-Based Insurance
The traditional insurance model often requires customers to purchase coverage for a fixed period, regardless of whether they need it continuously. On-demand insurance addresses this by offering flexible, short-term coverage options that can be activated as needed. This model is particularly appealing to the modern consumer, who values convenience and only wants to pay for insurance when necessary.

Development of On-Demand Insurance Products
On-demand insurance products provide coverage for specific situations or periods. Examples include travel insurance, event insurance, and gig economy insurance. These products cater to the growing demand for flexible and immediate coverage options.

Leveraging Generative AI and LLMs
Generative AI and LLMs streamline the creation and customization of on-demand insurance products. These technologies analyze customer data to predict when and what type of coverage might be needed, offering personalized insurance solutions. AI automates the activation and deactivation of policies based on real-time data, ensuring that customers are only charged for the coverage they use. LLMs generate clear and concise policy documents, simplifying the process for customers. For example, Trov, an on-demand insurance company, uses AI to offer flexible insurance coverage for personal items, allowing customers to turn coverage on or off via a mobile app.

7.7. Environmental and Climate Risk Insurance

Introduction to Climate Risk Insurance
As climate change continues to escalate, businesses and individuals face increasing risks from natural disasters such as floods, hurricanes, and wildfires. Traditional insurance products often do not provide adequate coverage for these evolving risks. Climate risk insurance products address the specific challenges posed by climate change, offering more comprehensive protection.

Development of Climate Risk Insurance Products
Climate risk insurance products include catastrophe insurance, parametric insurance, and agricultural insurance. These products provide coverage for large-scale natural

disasters, with parametric policies paying out based on predefined triggers rather than actual losses. Agricultural insurance offers protection against risks like drought, excessive rainfall, and pest infestations.

Leveraging Generative AI and LLMs

Generative AI and LLMs enhance the development of climate risk insurance by analyzing large datasets on weather patterns, historical claims, and environmental changes. These technologies help insurers model future climate scenarios and assess the potential impact on insured assets. AI facilitates the creation of parametric insurance products by identifying suitable triggers and setting appropriate payout thresholds. LLMs assist in communicating complex climate risk information to policyholders, helping them understand their coverage options and associated risks. Swiss Re, a leading reinsurance company, uses AI to develop advanced models for assessing climate risks and creating innovative insurance solutions.

Health and Wellness Insurance

Introduction to Personalized Health Insurance

Traditional health insurance often provides standardized coverage that does not account for individual health needs and behaviors. Personalized health insurance products offer tailored coverage and incentives for healthy living, aligning with the growing consumer focus on wellness and preventive care.

Development of Personalized Health Insurance Products

Personalized health insurance products include wellness programs, wearable-integrated insurance, and chronic disease management. Wellness programs offer rewards for engaging in healthy behaviors, while wearable-integrated insurance adjusts premiums based on real-time health data. Chronic disease management provides specialized coverage and support for individuals with chronic conditions.

Leveraging Generative AI and LLMs

Generative AI and LLMs significantly enhance personalized health insurance by analyzing individual health data to provide customized coverage recommendations and wellness programs. AI monitors data from wearable devices, predicts health risks, and suggests preventive measures, while LLMs generate personalized health advice and policy documents. Additionally, AI helps insurers design incentive structures that encourage healthy behaviors and improve overall health outcomes for policyholders. Vitality, a health and life insurance provider, uses AI to analyze data from wearable fitness trackers, offering personalized wellness programs and incentives based on individual health metrics.

CHAPTER 7   THE EVOLUTION OF INSURANCE IN THE DIGITAL AGE

Future of Insurance Product Innovation

The continuous integration of Generative AI and LLMs in insurance product development is set to drive further innovations. As these technologies evolve, they will enable insurers to create even more personalized, flexible, and comprehensive insurance solutions. By leveraging AI-driven insights and automated processes, insurers can better meet the diverse and dynamic needs of modern consumers, ensuring they remain relevant and competitive in a rapidly changing landscape.

Pioneering Product Innovations in Insurance

The traditional insurance product landscape has been largely static, relying on standardized policies designed to cover a broad spectrum of risks. However, the rapidly evolving risk environment and changing consumer expectations necessitate a shift towards more innovative insurance products. Today's consumers demand flexibility, personalization, and coverage for emerging risks that traditional policies do not adequately address. This section explores the emergence of new insurance products and services that cater to specific modern-day needs and preferences, leveraging cutting-edge technologies such as Generative AI and Large Language Models (LLMs) to enhance innovation.

Introducing On-Demand Insurance

On-demand insurance is a response to the modern consumer's desire for greater control over their insurance protection. Unlike traditional policies that require long-term commitments, on-demand insurance allows policyholders to activate and deactivate coverage as needed. This flexibility is particularly appealing to individuals who seek to manage their insurance costs more effectively and ensure they are only paying for coverage when necessary.

Examples of On-Demand Insurance

1. **Travel Insurance Activated by Geolocation:** Travelers can activate insurance coverage as soon as they leave their home country and deactivate it upon return. Geolocation technology ensures that coverage is automatically turned on when the policyholder enters a foreign country and turned off when they return home, eliminating the need for manual adjustments.

2. **Property Insurance for Shared Economy Participants:** Homeowners who rent out their properties on platforms like Airbnb can activate insurance coverage specifically for the duration of the rental period. This ensures that they are protected against risks associated with hosting guests without paying for continuous coverage.

Impact on Consumer Behavior and Insurance Purchasing Patterns

On-demand insurance has significantly influenced consumer behavior and insurance purchasing patterns. Consumers are now more likely to seek insurance solutions that offer flexibility and can be tailored to their specific needs. This shift has led to increased engagement with insurance providers and a greater willingness to purchase coverage for short-term or situational risks. However, on-demand insurance also presents challenges in underwriting and policy management. Insurers must develop sophisticated algorithms to assess risks accurately in real time and price policies accordingly. Additionally, managing a dynamic portfolio of on-demand policies requires advanced data analytics and automation to ensure efficiency and accuracy.

Usage-Based Insurance (UBI): Personalization and Fair Pricing

Principles Behind Usage-Based Insurance Models

Usage-Based Insurance (UBI) models, particularly prevalent in auto insurance, are designed to offer personalized and fair pricing based on actual usage or behavior. Unlike traditional insurance policies that rely on broad demographic factors, UBI assesses premiums based on how, when, and how much a policyholder uses their vehicle. This approach provides a more accurate assessment of risk and aligns insurance costs with actual driving behavior.

Technology Enabling UBI

Telematics devices and smartphone apps are the primary technologies enabling UBI. These devices collect data on various driving parameters, including mileage, driving behavior (speed, acceleration, braking), and driving conditions (time of day, weather). This data is transmitted to insurers, who use advanced analytics and machine learning algorithms to evaluate risk and determine personalized premiums.

Benefits for Insurers and Policyholders

UBI offers several benefits for both insurers and policyholders:

- **Incentivizing Safer Driving:** By directly linking premiums to driving behavior, UBI encourages policyholders to adopt safer driving habits to reduce their insurance costs. This can lead to fewer accidents and lower overall claims for insurers.

- **Potentially Lower Premium Costs:** Safe drivers can benefit from significantly reduced premiums compared to traditional insurance models, which often do not account for individual driving habits.

- **Enhanced Risk Assessment:** Insurers gain a more detailed and accurate understanding of each policyholder's risk profile, enabling them to price policies more fairly and competitively.

However, the use of telematics and continuous data collection raises privacy concerns. Policyholders may be wary of sharing detailed information about their driving habits and locations. Insurers must address these concerns by ensuring robust data security measures and transparent data usage policies.

Cyber Insurance: Safeguarding Digital Assets

Rise of Cyber Risks

With the increasing digitalization of business operations, cyber risks have become a significant concern. Data breaches, hacking, ransomware, and online fraud pose substantial threats to organizations, leading to financial losses, reputational damage, and operational disruptions. As these cyber threats proliferate, there is a growing demand for insurance products that offer protection against such risks.

Development of Cyber Insurance Policies

Cyber insurance policies have evolved to provide comprehensive coverage for a range of cyber-related incidents. These policies typically include coverage options such as data breach response, business interruption losses, cyber extortion, and liability coverage. The development and pricing of cyber insurance policies pose unique challenges. Unlike traditional risks, cyber risks are constantly evolving, making it difficult to predict and assess accurately. Insurers must consider factors such as business size, industry, cybersecurity measures, and data handled.

Leveraging Generative AI and LLMs

Generative AI and LLMs can significantly enhance the development and customization of cyber insurance products. By analyzing vast amounts of data on cyber incidents and their impacts, these technologies can help insurers identify patterns and predict future risks more accurately. LLMs assist in drafting tailored policy documents and generating personalized recommendations for businesses based on their specific risk profiles. AI-driven simulations can model potential cyberattack scenarios, helping insurers refine coverage options and pricing strategies. For instance, AXA has utilized AI to better understand cyber risks and tailor their insurance products accordingly.

Market Growth and Acceptance

The market for cyber insurance has seen substantial growth, driven by increased awareness of cyber risks and the need for robust protection. Businesses of all sizes recognize the importance of cyber insurance as part of their risk management strategy. According to industry reports, the global cyber insurance market is projected to grow significantly, with more companies seeking coverage as regulatory requirements and cybersecurity threats intensify. Case studies highlight the growing acceptance and reliance on cyber insurance. For example, a mid-sized company that suffered a ransomware attack leveraged its cyber insurance policy to cover the ransom payment, restore its systems, and manage the associated legal and reputational fallout. This real-world example underscores the value of cyber insurance in mitigating the financial impact of cyber incidents.

Emerging Risks and Innovative Coverage Solutions

In response to emerging risks, the insurance industry has developed new products that address specific modern-day challenges. These include climate change-related insurance products and health insurance that incorporates wellness and preventive care. Identifying new risks and developing corresponding insurance solutions involves complex actuarial modeling and navigating regulatory approval processes.

Climate Change-Related Insurance Products

With the increasing frequency and severity of natural disasters, climate change-related insurance products have become essential. These include

- **Catastrophe Insurance:** Protection against large-scale natural disasters, covering property damage, business interruption, and recovery costs

- **Parametric Insurance:** Policies that pay out based on predefined triggers, such as the magnitude of an earthquake or the amount of rainfall during a storm, rather than actual losses

- **Agricultural Insurance:** Coverage for farmers and agribusinesses against risks like drought, excessive rainfall, and pest infestations

Health Insurance Incorporating Wellness and Preventive Care

Personalized health insurance products focus on wellness and preventive care to improve overall health outcomes. These include

- **Wellness Programs:** Insurance plans that offer rewards and discounts for engaging in healthy behaviors, such as regular exercise, healthy eating, and routine medical check-ups

- **Wearable-Integrated Insurance:** Policies that integrate with wearable devices to monitor health metrics and adjust premiums based on real-time data

- **Chronic Disease Management:** Specialized coverage and support for individuals with chronic conditions, offering tailored treatment plans and medication management

Challenges in Identifying and Addressing New Risks

Developing innovative insurance solutions for emerging risks involves several challenges:

- **Actuarial Modeling:** Accurately predicting and pricing new risks requires advanced actuarial models that can handle large datasets and complex variables. Generative AI and LLMs can assist in building these models by analyzing historical data and simulating potential future scenarios.

- **Regulatory Approval:** Securing regulatory approval for new insurance products can be a lengthy and complex process. Insurers must work closely with regulatory bodies to ensure compliance with existing laws and advocate for the adaptation of regulations to accommodate innovative products.

Market Response and Consumer Adoption

Market Response to Innovative Insurance Products

The market response to innovative insurance products has been generally positive, with increasing adoption rates and favorable consumer feedback. Consumers appreciate the flexibility, personalization, and comprehensive coverage offered by these new products. For example:

- **On-Demand Insurance:** Adoption rates for on-demand insurance products have surged, particularly among younger consumers who value convenience and control. Insurers have reported increased engagement and higher customer satisfaction with these flexible offerings.

- **Usage-Based Insurance:** UBI has gained traction among drivers who benefit from personalized premiums based on their driving behavior. Studies have shown that UBI policyholders exhibit safer driving habits, leading to lower claims and reduced premiums.

- **Cyber Insurance:** The demand for cyber insurance has grown significantly as businesses recognize the importance of protecting digital assets. Increased awareness and regulatory requirements have driven higher uptake rates.

Role of Consumer Education and Marketing

Consumer education and marketing play a crucial role in promoting the uptake of new insurance models. Insurers must effectively communicate the benefits and features of innovative products to potential customers. Strategies include

- **Educational Campaigns:** Informing consumers about emerging risks and the importance of tailored insurance solutions through online content, webinars, and workshops

- **Targeted Marketing:** Using data analytics and AI to identify and target potential customers with personalized marketing messages that highlight the relevance and advantages of innovative insurance products

- **Transparency and Trust:** Building trust with consumers by being transparent about data usage, pricing models, and the value proposition of new insurance products

By addressing these aspects, insurers can foster greater acceptance and adoption of innovative insurance solutions, ensuring they meet the evolving needs of modern consumers while staying competitive in the dynamic insurance market.

Challenges in Innovation and Product Development

Insurers face several challenges in developing and launching new products. One major hurdle is navigating complex regulatory landscapes. Ensuring that innovative products comply with existing laws and regulations, which may not be designed to accommodate new types of coverage, can be time-consuming and challenging. For instance, regulatory bodies might not have clear guidelines for new insurance models like pay-as-you-go or dynamic pricing based on real-time data.

Another significant challenge is the investment required in technology and data analytics. Developing advanced insurance products necessitates substantial investments in robust IT infrastructure, hiring skilled data scientists, and continuously updating systems to handle large volumes of data. For example, implementing generative AI for personalized health insurance requires constant data integration from wearable devices and analysis.

Balancing innovation with effective risk management is also crucial. While introducing new products is essential for staying competitive, insurers must carefully assess and mitigate potential risks. New products often come with uncertainties that need thorough evaluation to avoid unforeseen liabilities.

Strategies for Overcoming Challenges

To overcome these challenges, insurers can adopt several strategies. Forming partnerships with technology firms and Insurtech startups can provide access to cutting-edge technologies and expertise. Collaborations like that of AXA with AI companies have accelerated product development and enhanced technological capabilities.

Implementing agile product development methodologies allows insurers to develop and launch products more quickly and efficiently. Agile practices, such as iterative development and continuous feedback, enable insurers to respond promptly to market changes and customer needs. Companies like Lemonade use agile approaches to innovate rapidly in the insurance space.

Proactively engaging with regulatory bodies can help insurers navigate regulatory hurdles and advocate for regulatory adaptations to support innovation. Open dialogue with regulators ensures new products meet compliance requirements and gain approval more smoothly. For instance, the collaboration between insurers and the New York Department of Financial Services (NYDFS) on cybersecurity regulations showcases proactive regulatory engagement.

The Future of Insurance Product Innovation

Several trends are expected to shape the future of insurance product innovation. Advances in technology, such as blockchain, Augmented Reality (AR), and advanced biometrics, will continue to revolutionize insurance products. Blockchain can enhance data security and transparency, AR can improve damage assessments, and biometrics can enhance identity verification processes. For example, Nationwide Insurance is exploring blockchain to streamline claims processing.

Changing consumer expectations will also drive the need for more tailored and flexible products. As consumers become more tech-savvy, insurers will need to leverage AI and data analytics to meet these expectations. Companies like Progressive use AI to offer personalized auto insurance rates based on driving behavior.

Emerging global risks, such as climate change, pandemics, and geopolitical instability, will necessitate innovative insurance solutions. Insurers must stay ahead of these risks by developing products that address evolving challenges. Swiss Re, for instance, uses AI to model climate risks and create innovative insurance products.

Staying Ahead of the Curve

To stay ahead of the curve, insurers can continuously invest in research and development (R&D). This proactive approach allows them to explore new technologies and identify emerging risks, leading to the creation of cutting-edge products that meet future market needs. Companies like Allstate invest heavily in R&D to drive innovation.

Fostering a culture of innovation within the organization is also crucial. Insurers should empower employees to propose new ideas and collaborate on innovative projects. Encouraging creativity and experimentation can drive meaningful innovation. USAA, for instance, promotes an innovative culture through internal hackathons and idea competitions.

Regularly engaging with consumers to understand their needs and preferences can inform product development. Feedback from policyholders helps insurers design relevant and valuable products. This engagement ensures that new products resonate with consumers and meet their expectations. MetLife, for example, uses customer feedback to refine its insurance offerings.

Navigating the Complex Regulatory Landscape

The insurance industry is undergoing significant transformation due to digital advancements, leading to heightened regulatory scrutiny and evolving compliance requirements. Insurers must navigate complex regulations governing data protection, cybersecurity, and digital customer interactions.

Key Regulatory Issues in Digital Insurance

Data protection and privacy are critical concerns in the digital insurance landscape (Table 7-3). Regulations such as the General Data Protection Regulation (GDPR) in Europe and the California Consumer Privacy Act (CCPA) in the United States impose stringent requirements on how insurers collect, store, and use personal data. Insurers like Allianz have implemented robust data management practices to comply with these regulations.

CHAPTER 7   THE EVOLUTION OF INSURANCE IN THE DIGITAL AGE

Table 7-3. Regulatory Challenges

Regulatory Challenges	Description	Key Aspects	Examples
Data Protection and Privacy Laws	Compliance with GDPR, CCPA, and similar laws that govern data collection, storage, and usage.	Consent, transparency, data minimization, rights of individuals.	Allianz implements robust data management practices to comply with GDPR and CCPA.
Cybersecurity Regulations	Implementing measures to protect sensitive data from breaches and cyberattacks, including compliance with NYDFS Cybersecurity Regulation.	Technical and organizational measures, risk assessments, incident reporting.	Chubb adopts advanced cybersecurity protocols to comply with NYDFS Cybersecurity Regulation.
Standards for Digital Customer Interaction	Ensuring clear disclosure of terms, obtaining informed consent, and making digital platforms accessible.	Disclosure requirements, customer consent, accessibility standards.	Prudential invests in user-friendly digital interfaces and transparent communication practices.
Global Variation in Regulatory Frameworks	Navigating diverse regulations across different countries, aligning business practices with local requirements.	Regulatory alignment, cross-border data transfers, localized compliance programs.	AIG develops a comprehensive global compliance strategy to navigate diverse regulatory frameworks.

Cybersecurity regulations are essential to protecting sensitive data from breaches and cyberattacks. Insurers must implement comprehensive cybersecurity measures to safeguard personal and financial information. The NYDFS Cybersecurity Regulation requires insurers to establish and maintain cybersecurity programs, conduct risk assessments, and report cybersecurity events. Insurers like Chubb have adopted advanced cybersecurity protocols to ensure compliance.

As insurers increasingly interact with customers through digital channels, they must comply with standards ensuring fair and transparent communication. This includes clear disclosure of terms, obtaining informed consent for digital communications, and ensuring digital platforms are accessible to individuals with disabilities. Insurers like Prudential have invested in user-friendly digital interfaces and transparent communication practices.

Global Variation in Regulatory Frameworks
Multinational insurance operations face the challenge of navigating diverse regulatory frameworks across different regions. Each country may have unique regulations governing data protection, cybersecurity, and digital interactions, complicating compliance efforts and increasing operational costs. Insurers must ensure regulatory alignment, comply with cross-border data transfer regulations, and develop localized compliance programs. For example, AIG has developed a comprehensive global compliance strategy to address these challenges.

By embracing these strategies and addressing the challenges head-on, insurers can remain competitive and responsive to the rapidly changing landscape, ensuring they continue to meet the diverse and dynamic needs of their customers.

7.8. The Impact of Digital Transformation on Compliance

Introduction to Digital Transformation Challenges
Digital transformation initiatives, such as the adoption of cloud computing, AI, and IoT devices, introduce new compliance challenges for insurers. These technologies offer numerous benefits but also require insurers to navigate complex regulatory landscapes and ensure compliance with evolving standards.

Cloud Computing
Cloud computing enables insurers to store and process large volumes of data efficiently. However, it also raises concerns about data security, privacy, and regulatory compliance. Key regulatory considerations include

- **Data Sovereignty:** Ensuring that data stored in the cloud complies with local data protection laws, which may require data to be stored within specific geographic boundaries.

- **Access Controls:** Implementing robust access controls to prevent unauthorized access to sensitive data stored in the cloud.

- **Third-Party Risk Management:** Assessing and managing risks associated with third-party cloud service providers, ensuring they comply with relevant regulations.

For example, Allianz uses cloud computing to enhance its data processing capabilities while implementing stringent access controls and compliance checks to meet data sovereignty requirements.

Artificial Intelligence (AI)

AI technologies, such as machine learning and predictive analytics, can enhance underwriting, claims processing, and customer service. However, they also introduce challenges related to transparency, bias, and accountability. Key regulatory considerations include

- **Algorithmic Transparency:** Ensuring that AI algorithms are transparent and explainable, allowing regulators and stakeholders to understand how decisions are made.

- **Bias and Fairness:** Implementing measures to detect and mitigate bias in AI models to ensure fair and non-discriminatory outcomes.

- **Accountability:** Establishing clear accountability for AI-driven decisions and ensuring compliance with ethical standards.

For instance, AXA uses AI to streamline claims processing while ensuring algorithmic transparency and bias mitigation through continuous monitoring and adjustments.

The Role of Regulatory Technology (RegTech)

Regulatory Technology (RegTech) leverages automation and data analytics to streamline regulatory compliance processes. RegTech solutions help insurers manage compliance more efficiently by automating routine tasks, monitoring regulatory changes, and conducting risk assessments.

Benefits of RegTech Solutions

RegTech solutions offer several benefits for insurers, including

- **Automated Compliance Monitoring:** RegTech tools can automatically monitor and analyze regulatory changes, ensuring that insurers stay up-to-date with evolving requirements. This reduces the risk of non-compliance and associated penalties.

- **Compliance Reporting:** Automated compliance reporting tools streamline the process of generating and submitting reports to regulatory authorities, reducing the administrative burden on compliance teams.

- **Risk Assessments:** Advanced analytics and machine learning algorithms enable RegTech solutions to conduct real-time risk assessments, identifying potential compliance issues before they become critical problems.

- **Data Management:** RegTech platforms can integrate with insurers' existing systems to manage and analyze large volumes of data, ensuring that compliance-related data is accurate, secure, and readily accessible.

For instance, Lloyd's of London uses RegTech to enhance compliance monitoring and reporting, ensuring they meet regulatory standards efficiently.

Implementing RegTech Solutions

To effectively implement RegTech solutions, insurers can follow these strategies:

- **Integration with Existing Systems:** Ensure that RegTech solutions seamlessly integrate with existing IT infrastructure to enable smooth data flow and minimize disruptions.

- **Collaboration with RegTech Providers:** Partner with experienced RegTech providers who understand the regulatory landscape and can offer tailored solutions.

- **Employee Training:** Invest in training programs to ensure that employees are proficient in using RegTech tools and understand their role in the compliance process.

- **Continuous Improvement:** Regularly review and update RegTech solutions to adapt to changing regulatory requirements and technological advancements.

By leveraging RegTech, insurers can enhance their compliance capabilities, reduce operational costs, and focus on innovation and growth in a rapidly evolving regulatory environment.

Strategies for Navigating Regulatory Changes

Proactive Engagement with Regulators

Maintaining proactive engagement with regulatory bodies is crucial for insurers to stay ahead of regulatory changes. Strategies include

- **Regular Dialogue:** Establishing regular communication channels with regulators to discuss upcoming regulations, industry trends, and potential compliance issues.

- **Participating in Consultations:** Actively participating in regulatory consultations and public comment periods to provide feedback and insights on proposed regulations.

- **Industry Associations:** Joining industry associations and working groups to stay informed about regulatory developments and advocate for favorable regulatory environments.

Investing in Compliance Training for Staff

Continuous training and development programs are essential to ensure that staff members are knowledgeable about regulatory requirements and best practices. Strategies include

- **Ongoing Education:** Providing regular training sessions and workshops on compliance topics, including data protection, cybersecurity, and ethical AI use.

- **Certifications and Courses:** Encouraging staff to pursue relevant certifications and courses to deepen their understanding of regulatory issues.

- **Compliance Culture:** Fostering a culture of compliance within the organization by emphasizing the importance of ethical behavior and regulatory adherence.

Integrating Compliance Considerations into Product Development

Integrating compliance considerations into the product development process helps insurers design products that meet regulatory requirements from the outset. Strategies include

- **Compliance by Design:** Incorporating regulatory requirements into the product design and development phases to ensure that new products comply with relevant laws and standards.

- **Cross-Functional Teams:** Establishing cross-functional teams that include compliance experts, legal advisors, and product developers to collaborate on new product initiatives.

- **Regular Audits and Reviews:** Conducting regular audits and reviews of new products to identify and address potential compliance issues before launch.

Importance of Regulatory Foresight and Agility

Regulatory foresight and agility are critical for insurers to adapt to new regulatory requirements effectively. Strategies include

- **Regulatory Forecasting:** Leveraging data analytics and forecasting tools to anticipate future regulatory trends and prepare for potential changes.

- **Agile Business Models:** Adopting agile business models that allow for rapid adjustments to operations, products, and processes in response to regulatory developments.

- **Scenario Planning:** Conducting scenario planning exercises to assess the potential impact of regulatory changes and develop contingency plans.

Collaboration Between Regulators and the Insurance Industry

Collaboration and open dialogue between insurance companies and regulatory bodies foster a better understanding of digital transformation challenges and help create a more supportive regulatory environment. Engaging with regulators can lead to more informed and balanced regulatory frameworks that accommodate technological advancements while ensuring consumer protection.

By addressing these regulatory challenges through proactive strategies and leveraging technology, insurers can navigate the complex compliance landscape, enabling them to innovate and grow in a rapidly evolving market.

In conclusion, the insurance industry is undergoing a profound transformation driven by digital innovation and evolving consumer demands. The integration of technologies such as cloud computing, AI, and IoT devices has introduced new compliance challenges but also provided opportunities for insurers to enhance their offerings. By leveraging Generative AI and Large Language Models (LLMs), insurers can develop more personalized and flexible products that meet the specific needs of modern consumers. Addressing regulatory hurdles, investing in technology, and fostering a culture of innovation are crucial for staying competitive. As insurers navigate this dynamic landscape, proactive engagement with regulators and continuous adaptation to emerging risks will be essential in shaping the future of the industry.

7.9. Summary

- The traditional insurance landscape has been largely static, relying on standardized policies that fail to address the emerging risks and changing consumer expectations in the digital age.

- The integration of Generative AI and Large Language Models (LLMs) is revolutionizing the insurance industry by enabling the development of innovative, personalized products that cater to specific modern-day needs.

- Digital transformation is reshaping the insurance sector by improving operational efficiencies, automating claims processing, and enhancing customer engagement through AI-driven tools such as chatbots and virtual assistants.

- The use of IoT devices in insurance allows for real-time data collection, enabling insurers to create more accurate and dynamic pricing models based on individual behavior and risk profiles.

- Cyber insurance is becoming increasingly important as digitalization exposes businesses to significant risks like data breaches and cyberattacks. AI is enhancing the development of these policies by providing detailed risk assessments and tailored coverage options.

CHAPTER 7   THE EVOLUTION OF INSURANCE IN THE DIGITAL AGE

- On-demand and usage-based insurance models are gaining popularity, offering consumers more flexibility and personalized pricing based on actual usage or behavior, which is particularly appealing to modern, tech-savvy customers.

- AI and machine learning are transforming customer engagement by providing personalized recommendations and improving customer experiences through more responsive and tailored interactions.

- Insurers face challenges in data privacy, regulatory compliance, and technological investment as they navigate the complex landscape of digital transformation, requiring proactive strategies to ensure security and adherence to regulations.

- Blockchain technology offers significant potential for enhancing data integrity and transparency in insurance, particularly in scenarios requiring secure data sharing and fraud prevention.

- The future of insurance product innovation lies in the continuous integration of AI and emerging technologies, which will drive the creation of more personalized, flexible, and comprehensive insurance solutions to meet evolving consumer needs.

PART III

The Road Ahead for Generative AI in BFSI

CHAPTER 8

Roadmap for AI Implementation in BFSI

In the Banking, Financial Services, and Insurance (BFSI) sector, rolling out artificial intelligence (AI) demands a thoughtful and well-planned approach, starting with an exhaustive evaluation of the current tech setup. This vital initial step involves thoroughly exploring the existing technological framework (Figure 8-1) and deeply understanding the distinctive requirements and obstacles inherent in the BFSI field. Such an evaluation not only sheds light on the present conditions but also pinpoints where AI can be integrated effectively. Building upon this groundwork, the next essential phase is developing a strategy that dovetails with the organization's broader business goals. This plan must tackle the unique challenges anticipated in AI adoption, which range from integrating new technologies to navigating regulatory landscapes and addressing ethical issues. With meticulous planning and forward-thinking, BFSI entities can establish a route that utilizes AI to bolster operational efficiency, enrich customer interactions, and sustain a competitive advantage in a progressively digitized market environment.

Chapter 8 Roadmap for AI Implementation in BFSI

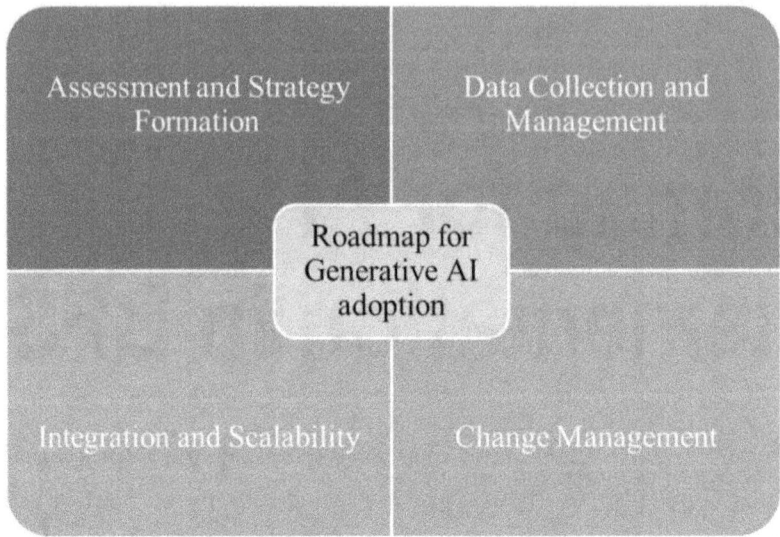

Figure 8-1. *Generative AI Roadmap*

8.1. Assessment and Strategy Formation

In the BFSI sector, initiating artificial intelligence (AI) integration starts with a deep evaluation of the current technological infrastructure. This foundational step involves a detailed review of the systems and methods in place, aiming to spot deficiencies or areas where the setup does not meet industry standards. By comparing these findings to the best practices of the industry, organizations can pinpoint exact areas for significant enhancements through AI, which transcend mere incremental upgrades.

Progressing to the assessment of BFSI-specific needs, it's vital to consider the unique challenges and strict regulatory requirements inherent in this sector. This phase should include a thorough analysis of regulatory frameworks, customer expectations, and the essential need for robust data security. Understanding these specific requirements ensures that AI solutions are crafted not only to boost operational efficiencies but also to comply with strict data protection norms and ethical standards, thereby fostering AI applications that tackle the sector's unique challenges alongside aiming for higher customer engagement.

For effective AI integration in the BFSI sector, aligning AI with business objectives is crucial. This alignment entails setting clear goals for AI deployment and ensuring these initiatives are in harmony with the organization's overall business strategy. Identifying and prioritizing AI projects based on their potential business impact helps ensure that

AI investments are driven by more than the allure of new tech but are chosen for their capacity to significantly advance the organization's strategic aims, thus supporting sustainable growth and a competitive edge.

Moreover, integrating an overarching organizational AI policy and a clear AI strategy is crucial. Establishing a comprehensive governance framework to ensure adherence to this policy is fundamental. This framework will guide the ethical development, deployment, and maintenance of AI systems, ensuring that all AI initiatives are aligned with both the organization's values and regulatory demands. By embedding these governance structures, organizations can ensure that AI technologies are used responsibly and effectively across all levels, enhancing trust and consistency in AI applications.

Navigating the initial steps of technology assessment, BFSI-specific evaluation, and strategic alignment and progressing through robust governance implementation prepares organizations to address AI integration challenges effectively. This structured approach not only facilitates the successful adoption of AI but also ensures that the deployment of these technologies is purposeful and impactful, driving the sector towards greater innovation and excellence in service.

8.2. Data Collection and Management

In the dynamic world of artificial intelligence (AI) within the Banking, Financial Services, and Insurance (BFSI) sector, data is the linchpin of all AI operations. Its role is crucial—it's the lifeblood that powers algorithms responsible for decision-making, risk evaluations, customer interactions, and tailored services. It's essential that data collection adheres to the highest standards, ensuring that the data not only covers a broad spectrum but is also precise, complete, and consistent. Effective data management strategies are key, organizing and leveraging this precious resource to fuel innovation and sharpen competitive edges.

Consider the example of JPMorgan Chase, which uses AI-driven data analytics to optimize trading strategies, resulting in significant profit margins. In contrast, Wells Fargo faced severe reputational damage and financial penalties due to poor data management practices that led to customer account fraud. These cases underscore the impact of data quality on AI outcomes.

However, amidst this enthusiasm for data's possibilities, the imperative of data privacy and security cannot be ignored. Given the delicate nature of financial information, BFSI organizations face the dual challenge of maximizing data's utility for AI advancements while rigorously protecting it. They must ensure robust security measures to prevent data breaches, comply with strict regulatory frameworks, and sustain customer trust, which is fundamental in the financial sector.

At the core of any AI system, especially in the BFSI sector, the axiom holds: the quality and volume of the data (Table 8-1) it processes directly influence its efficacy and reliability. Data here transcends its role as a mere operational tool; it becomes a strategic asset that informs decision-making, enriches customer experiences, and streamlines operations. The stakes of data quality are high—excellent, timely data can propel accurate AI forecasts and decisions, whereas subpar data can mislead, leading to substantial financial and reputational repercussions. Thus, viewing data as a strategic asset underscores the necessity for stringent data governance and management, ensuring its integrity, applicability, and security at every turn.

Table 8-1. Data Collection and Management

Data Collection	Data Management
– Relevance and Sources	– Quality Assurance
– Traditional Data Pools	– Accuracy, Completeness, Consistency
– Unconventional Sources	– Data Cleaning and Preprocessing
– Advanced Data Extraction Tools and APIs	– Regular Audits and Quality Checks
– Bias and Representativeness	– Strategies
– Ensuring Diverse Data	– Storage and Organization
– Minimizing Bias	– Data Governance Frameworks
	– Lifecycle Management
– Regulatory Compliance	– Ethical Considerations
– GDPR, CCPA	– Bias, Transparency, Accountability
– Strict Regulatory Frameworks	– Ethical Policies
	– Future Trends

By addressing these critical aspects—high-quality data collection, rigorous data management, and robust data privacy—BFSI organizations can navigate the complexities of AI implementation more effectively. This comprehensive approach ensures that AI systems deliver meaningful insights and drive the sector towards greater innovation and service excellence.

Moving on to best practices in data collection, identifying relevant data sources is crucial. In the BFSI sector, this involves tapping into both traditional data pools and exploring unconventional sources to gain unique insights and a competitive edge. Advanced data extraction tools, APIs for real-time data collection, and partnerships with data providers are essential for efficient and comprehensive data gathering. This ensures that the data collected is extensive, relevant, and actionable. Ensuring representativeness and minimizing bias in data collection is another critical aspect. Deliberate actions must be taken to guarantee that the data reflects the diversity of the customer base and the operational environments within which the BFSI sector operates. Addressing these aspects from the outset significantly reduces the risk of embedding biases into AI systems, leading to more equitable and accurate outcomes.

Most BFSI organizations have vetted models for data storage, with data collection governed by strict regulatory frameworks. This data is not only a critical asset but also unique to each customer. However, current Generative AI (GenAI) use cases often rely on readily available Large Language Models (LLMs) or sometimes Small Language Models (SLMs). In these scenarios, BFSI customers typically do not share their proprietary data with model providers. Instead, their data is predominantly used for grounding or fine-tuning the models. Given this context, many considerations for training LLMs do not directly apply to BFSI customers. However, transparency about data collection techniques from model providers is crucial. BFSI organizations need disclosures to ensure adherence to their own AI policies and regulatory requirements. This is an evolving space, expected to mature over time.

Case Study: Transparency in Data Collection for GenAI at HSBC
HSBC, a global leader in the BFSI sector, integrates Generative AI using external Large Language Models (LLMs) without sharing proprietary customer data directly with model providers. Instead, HSBC uses customer data for grounding and fine-tuning these models. To ensure alignment with regulatory requirements and internal AI policies, HSBC mandates transparency from model providers regarding their data collection techniques. This approach allows HSBC to maintain strict data governance while leveraging advanced AI technologies, ensuring compliance and building trust with customers in an evolving AI landscape. This case highlights the importance of transparency in the BFSI sector's AI integration.

Data Quality Assurance is crucial for AI success in the Banking, Financial Services, and Insurance (BFSI) sector. High-quality data must be accurate, complete, and consistent. Accuracy ensures data reflects true values, aiding reliable decision-making. Completeness means no missing values, preventing incorrect AI predictions.

Consistency ensures data integrity across systems, avoiding discrepancies. Techniques for maintaining these standards include data cleaning and preprocessing. These involve correcting errors, filling missing values, and standardizing formats. For example, a bank might use automated tools to fix customer data discrepancies, ensuring AI applications are reliable. Regular audits and quality checks are essential. Setting up protocols to review data systematically keeps it high-quality and fit for use. Insurance companies, for instance, might conduct quarterly audits to verify claim records, crucial for reliable AI-driven assessments. Economically, high-quality data leads to cost savings and efficiency. Accurate data improves risk management and fraud detection, reducing losses. Socially, it ensures fair outcomes, enhancing trust and satisfaction. Ethically, it upholds transparency and accountability. Challenges include bias in data collection and processing, leading to skewed AI outcomes. Addressing this requires strict data governance and diverse data sources for representativeness. BFSI organizations must stay updated with regulatory compliance to avoid legal issues.

Effective data management strategies are essential for maximizing the value of data in the BFSI sector. Proper data storage and organization ensure that data is easily accessible, understandable, and analyzable by authorized personnel, facilitating swift and informed decision-making. Implementing data governance frameworks is crucial. These frameworks establish policies, procedures, and standards for managing data securely and ethically throughout its lifecycle. They also define roles and responsibilities, ensuring that those involved in data management understand their obligations regarding data quality, privacy, and protection.

Data lifecycle management in BFSI involves handling data from its creation to its eventual disposal or archiving. This process ensures that data remains accurate, accessible, and secure throughout its lifecycle. For example, a bank might implement strict protocols for data acquisition, usage, maintenance, and disposal to ensure compliance with regulatory standards and maintain customer trust. Additionally, integrating advanced data management tools can enhance data quality and security. For instance, using automated data validation tools can help identify and correct errors in real time, ensuring data integrity. Employing encryption and other security measures protects sensitive information from breaches, aligning with industry regulations and standards. From an economic perspective, effective data management reduces costs and enhances operational efficiency. Accurate and well-managed data improves decision-making and risk management, leading to better financial outcomes. Socially, it ensures fair and transparent practices, fostering customer trust and satisfaction.

Ethically, robust data management upholds principles of transparency, accountability, and privacy, essential in the BFSI sector. Challenges include maintaining data quality, ensuring regulatory compliance, and protecting against data breaches. Addressing these challenges requires continuous monitoring, regular audits, and updates to data management practices. By prioritizing effective data management, BFSI organizations can harness the full potential of their data, driving innovation and maintaining a competitive edge.

8.3. Integration and Scalability

Integrating and Scaling Generative AI in the BFSI Sector
Integrating generative AI into the Banking, Financial Services, and Insurance (BFSI) sector involves a complex but rewarding process of merging advanced AI technologies with existing legacy systems. This integration is akin to installing modern software on vintage hardware, necessitating careful calibration and customization. Generative AI can significantly enhance decision-making, customer service, and operational efficiency, but achieving these benefits requires a strategic approach to integration and scalability.

Strategic Integration Approach
To ensure a smooth integration, BFSI institutions must adopt a strategic approach that includes

1. **Assessment of Current Infrastructure:** Evaluating existing systems to identify compatibility issues and areas needing enhancement

2. **Incremental Implementation:** Introducing AI in phases to minimize disruption and allow for continuous improvement and adaptation

3. **Custom Middleware Development:** Creating custom middleware solutions to facilitate communication between legacy systems and AI technologies

Scalability Considerations

Scalability is critical for the long-term success of AI initiatives in the BFSI sector. AI solutions must not only integrate well but also scale efficiently as the business grows and evolves. This involves

1. **Scalable Architecture**: Designing AI systems with scalability in mind, using modular components that can be easily expanded or upgraded

2. **Cloud Integration**: Leveraging cloud computing to provide the flexibility and resources needed for scaling AI applications

3. **Continuous Monitoring and Optimization**: Implementing systems to monitor AI performance and make necessary adjustments to ensure optimal scalability

Organizational Adaptation

Successfully integrating and scaling generative AI requires an adaptive organizational culture. BFSI institutions must foster a culture of continuous learning and innovation, encouraging employees to embrace new technologies and methodologies. This involves

1. **Training and Upskilling**: Providing ongoing training programs to equip employees with the skills needed to work effectively with AI technologies

2. **Change Management**: Implementing change management strategies to help staff adapt to new workflows and processes introduced by AI integration

3. **Collaboration and Innovation**: Encouraging cross-functional collaboration to leverage diverse expertise and drive innovation

Technical Scalability

Technical scalability is crucial for AI systems in the BFSI sector, which must handle growing volumes of data and transactions. AI systems need to be designed with a forward-looking approach, ensuring they can expand and refine without complete overhauls. Building resilient and adaptable AI technologies that can scale as data volume increases ensures these solutions remain valuable as the business landscape evolves. This future-proofing of AI investments supports ongoing growth and transformation in an ever-changing market.

Organizational Scalability

For BFSI organizations to fully leverage AI, both systems and organizational structures must be prepared for AI-driven processes. This involves revising traditional models to embrace more agile, AI-compatible operations. Organizational scalability requires cultivating an AI-ready culture that encourages innovation and flexibility. Creating an environment where employees are open to change and continuous learning can smooth the transition to AI-centric operations, ensuring the workforce can effectively operate alongside advanced AI technologies.

Strategies for Successful AI Integration

1. **Flexible and Modular AI Solutions:**
   - Adopting AI solutions that work seamlessly with existing systems to minimize disruption and maximize innovation and efficiency gains.

2. **Continuous Training and Development:**
   - Investing in training programs tailored to the nuances of AI in the BFSI context. This empowers employees to work effectively with new technologies, enhancing both personal growth and the organization's competitive edge.

3. **Forward-Looking Planning:**
   - Implementing a forward-looking approach that considers immediate operational needs, long-term business objectives, and emerging technology trends. Continuous monitoring and adaptation of AI strategies ensure alignment with organizational goals and market changes.

Challenges and Opportunities in Integration

The BFSI sector faces unique challenges when integrating generative AI. Legacy systems, often not designed with AI in mind, present significant hurdles. These older systems may lack the flexibility and interoperability needed to seamlessly incorporate AI technologies. Overcoming these challenges requires innovative solutions, such as middleware that bridges the gap between old and new systems or the gradual phasing out of outdated components in favor of AI-compatible infrastructure.

Key Questions for Management to Consider When Implementing Generative AI

1. **Integration Readiness:**
   - How compatible are our current systems with generative AI technologies?
   - What steps are needed to bridge the gap between legacy systems and AI?

2. **Scalability:**
   - Can our AI solutions scale with increasing data volumes and transaction complexity?
   - What infrastructure changes are necessary to support scalable AI?

3. **Organizational Adaptation:**
   - Is our organizational structure flexible enough to accommodate AI-driven processes?
   - How can we cultivate a culture of continuous learning and innovation?

4. **Training and Development:**
   - What specific skills do our employees need to work effectively with generative AI?
   - How can we implement ongoing training programs to keep pace with AI advancements?

5. **Strategic Alignment:**
   - Are our AI initiatives aligned with our long-term business objectives?
   - How do we ensure our AI strategies remain adaptable to evolving market trends?

6. **Ethical Considerations:**
   - How do we address ethical concerns related to AI, such as bias and transparency?
   - What measures are in place to ensure ethical AI practices within our organization?

7. **Performance Monitoring:**
   - What metrics will we use to evaluate the success of our AI implementations?
   - How do we continuously monitor and optimize AI performance to meet our goals?

8.4. Change Management for AI Adoption in BFSI

As the Banking, Financial Services, and Insurance (BFSI) sector embarks on the journey of integrating artificial intelligence (AI) into its operations, the importance of a well-structured change management strategy cannot be overstated. This process, akin to preparing a vessel for a sea voyage, involves not only ensuring the ship is seaworthy but also that the crew is ready and the passengers are informed about the journey ahead. Within this context, change management splits into two main streams: internal and external. Internally, the focus is on cultivating an organizational culture that embraces AI through targeted training and active engagement with employees. Externally, the task involves managing the expectations of customers and navigating the complex waters of regulatory compliance, ensuring that all stakeholders are aligned and supportive of the AI transition. This dual approach, when executed thoughtfully, promises to smooth the path for AI adoption, making the journey less about confronting waves of resistance and more about sailing toward a horizon of innovation and efficiency.

Navigating the Integration of Generative AI in the BFSI Sector

In the ever-evolving landscape of the Banking, Financial Services, and Insurance (BFSI) sector, integrating Generative AI is a beacon of innovation, promising to reshape financial services delivery and management. However, fully embracing AI technology involves overcoming significant challenges, necessitating a well-orchestrated change management strategy to navigate the complexities involved. This process is akin to preparing for a major expedition; it requires meticulous planning, provisioning, and readiness to adapt to unforeseen obstacles along the way.

CHAPTER 8   ROADMAP FOR AI IMPLEMENTATION IN BFSI

The importance of comprehensive change management in successfully integrating Generative AI within the BFSI sector cannot be overstated. Similar to a ship's captain ensuring that every part of the vessel functions harmoniously, leaders in the BFSI sector must ensure that every aspect of their organization aligns with the new technological direction. This alignment goes beyond installing new software or deploying AI models; it involves transforming the organizational culture, processes, and mindset to fully leverage AI's benefits. Effective change management acts as the keel that keeps the ship steady, facilitating a smoother transition by addressing resistance, fostering acceptance, and ensuring the workforce is equipped and ready for the changes AI brings.

Internally, change management focuses on the organization's core—its people. Creating a culture that embraces change and views Generative AI as a tool for empowerment is essential. This cultural shift lays the foundation for successful AI adoption. Training and upskilling become key elements, ensuring employees are not left behind as the organization advances. For instance, a bank implementing an AI-driven customer service platform needs its staff to understand how this technology enhances their roles, equipping them with skills to manage complex customer needs that AI cannot address.

Externally, change management involves managing expectations and ensuring compliance with customers and regulatory bodies. For customers, this might involve educating them on how AI improves service delivery, such as through personalized banking advice or enhanced security measures. Regulatory compliance requires navigating complex laws and standards governing AI use, ensuring implementations are transparent, ethical, and in customers' best interests.

Complementary Roles of Internal and External Change Management
The roles of internal and external change management are complementary and crucial to AI adoption. Internally, the focus is on building a robust technological and cultural infrastructure to support AI integration. Externally, the emphasis shifts to creating an environment conducive to integration, where customers feel informed and protected, and regulatory bodies are assured of the technology's ethical application.

Key Questions for Management When Implementing Generative AI

1. **Integration Readiness:**
    - How compatible are our current systems with generative AI technologies?
    - What steps are needed to bridge the gap between legacy systems and AI?

2. **Scalability:**
   - Can our AI solutions scale with increasing data volumes and transaction complexity?
   - What infrastructure changes are necessary to support scalable AI?

3. **Organizational Adaptation:**
   - Is our organizational structure flexible enough to accommodate AI-driven processes?
   - How can we cultivate a culture of continuous learning and innovation?

4. **Training and Development:**
   - What specific skills do our employees need to work effectively with generative AI?
   - How can we implement ongoing training programs to keep pace with AI advancements?

5. **Strategic Alignment:**
   - Are our AI initiatives aligned with our long-term business objectives?
   - How do we ensure our AI strategies remain adaptable to evolving market trends?

6. **Ethical Considerations:**
   - How do we address ethical concerns related to AI, such as bias and transparency?
   - What measures are in place to ensure ethical AI practices within our organization?

7. **Performance Monitoring:**
   - What metrics will we use to evaluate the success of our AI implementations?
   - How do we continuously monitor and optimize AI performance to meet our goals?

Internal Change Management

Internal Change Management for Generative AI Adoption in the BFSI Sector

Internal change management is crucial for the successful adoption of Generative AI within the Banking, Financial Services, and Insurance (BFSI) sector. This transformation requires not only the integration of new technologies but also a significant shift in organizational culture, strategies for developing an AI-ready workforce, and mechanisms to ensure employees are engaged and informed throughout the process.

Cultural Shifts for AI Adoption

Creating a culture that embraces Generative AI is like preparing fertile ground for planting. It involves nurturing an environment where innovation can thrive. This cultural shift requires moving away from traditional ways of working and embracing AI-driven approaches. Leadership buy-in is essential, as leaders set the tone and direction for AI adoption. They act as role models by actively engaging with AI initiatives and demonstrating the value of these transformations to their teams.

Training and Development Programs

For Generative AI to be effectively integrated into BFSI operations, employees must understand how to work with these new technologies. Effective training programs should cover both the technical aspects of AI and its specific applications within the BFSI sector. Tailoring training to various roles ensures that each team understands how AI impacts their work. For instance, customer service representatives might receive training on AI chatbots, while analysts might focus on AI-driven data analysis tools.

Employee Engagement and Communication

Keeping employees engaged throughout the AI adoption process is critical. Engagement can be fostered through workshops, forums, and regular updates on AI projects. Transparent communication is vital, as it helps address concerns, manage expectations, and highlight the benefits of AI. Sharing success stories of how AI has improved processes or created new opportunities can illustrate the tangible benefits of AI adoption, helping to reduce resistance to change.

Checklist for Implementing Generative AI in the BFSI Sector

1. **Cultural Shift:**
    - Have leaders demonstrated a commitment to AI?
    - Are leaders actively engaging with AI initiatives and modeling desired behaviors?

2. **Training and Development:**
   - Are training programs tailored to specific roles and departments?
   - Do training programs cover both technical aspects and practical applications of AI within the BFSI sector?

3. **Employee Engagement:**
   - Are there regular updates on AI projects and their impact?
   - Are workshops and forums conducted to engage employees and gather feedback?

4. **Communication:**
   - Is there transparent communication about AI adoption, addressing concerns, and managing expectations?
   - Are success stories and tangible benefits of AI shared with employees?

External Change Management for Generative AI Adoption in the BFSI Sector

External change management is as critical as internal preparation when adopting Generative AI in the Banking, Financial Services, and Insurance (BFSI) sector. This aspect focuses on aligning customer expectations, navigating the regulatory landscape, and fostering collaborative partnerships to ensure a seamless transition to AI-enhanced operations.

Managing Customer Expectations and Experience

Communicating AI adoption to customers requires balancing the benefits and addressing potential concerns. Customers need to understand how AI will enhance services, such as through faster processing times, improved security measures, and personalized financial advice. It is equally important to address any apprehensions regarding data privacy and security. Clear and consistent communication through various channels—social media, email newsletters, and in-branch information sessions—can help manage expectations and reassure customers about the positive changes AI brings. Ensuring excellent customer service during this transition is vital; frontline staff should be well-equipped to answer questions about AI and maintain high service levels as AI technologies are integrated.

CHAPTER 8   ROADMAP FOR AI IMPLEMENTATION IN BFSI

Regulatory Compliance and Stakeholder Communication

Navigating the regulatory landscape associated with AI technologies is a significant aspect of external change management. The BFSI sector operates within stringent regulatory frameworks, making compliance crucial. Staying informed about current and upcoming regulations and engaging with regulatory bodies early in the AI adoption process is essential. Regular communication with regulators and other stakeholders ensures compliance and builds trust, demonstrating the organization's commitment to ethical AI use. This proactive approach can also provide insights into regulatory perspectives on AI, enabling more informed decision-making and strategy development.

Partnerships and Ecosystem Collaboration

Leveraging partnerships with technology providers, academic institutions, and industry consortia is crucial for supporting AI adoption. These collaborations offer access to the latest AI technologies, research findings, and best practices, enriching the organization's AI initiatives. Engaging with a broader ecosystem of AI practitioners allows for sharing learnings and challenges, contributing to collective advancements in AI technology. Such partnerships enhance the organization's AI capabilities and strengthen the industry's overall position at the technological forefront.

Checklist for Implementing Generative AI in BFSI Sector

1. **Managing Customer Expectations:**
    - Are customers informed about the benefits of AI, such as faster processing times and enhanced security?
    - Are there clear communication channels (social media, newsletters, in-branch sessions) to address customer concerns and expectations?

2. **Customer Service Excellence:**
    - Are frontline staff trained to answer questions about AI and maintain high service levels during the transition?

3. **Regulatory Compliance:**
    - Is the organization staying informed about current and upcoming AI regulations?
    - Are there regular communications with regulatory bodies to ensure compliance and build trust?

4. **Stakeholder Communication:**
   - Is there a proactive approach to engaging with stakeholders, demonstrating commitment to ethical AI use?

5. **Partnerships and Collaborations:**
   - Are there partnerships with technology providers, academic institutions, and industry consortia to support AI adoption?
   - Is the organization engaged in a broader AI ecosystem to share learnings and advancements?

8.5. Continuous Monitoring and Feedback Loops in AI Systems for BFSI

In the dynamic world of Banking, Financial Services, and Insurance (BFSI), the integration and evolution of Generative AI systems are essential for maintaining competitiveness and meeting customer needs. Ensuring these AI systems remain effective and up-to-date involves implementing continuous monitoring and feedback loops. These processes allow for the ongoing evaluation of AI applications, ensuring they perform as expected and adapt over time.

Case study (fictional): Capital One has effectively utilized continuous monitoring and feedback loops to keep its Generative AI systems in the BFSI sector competitive and adaptive. By regularly evaluating AI performance across areas like personalized financial advice and fraud detection, the company ensures its systems remain accurate, relevant, and secure. Real-time data streams and automated feedback loops allow the AI models to learn and adjust to changing customer behaviors and market conditions. This approach has enhanced customer satisfaction, improved fraud detection accuracy, and reinforced trust, demonstrating the critical role of ongoing monitoring in sustaining AI effectiveness in a dynamic environment.

Continuous Monitoring

Continuous monitoring is crucial for maintaining AI system integrity. It involves regularly observing system operations to ensure they run smoothly and efficiently, identifying deviations from expected performance early on. Key Performance Indicators (KPIs) and metrics, such as processing accuracy, response times, and user satisfaction rates, provide quantifiable measures of AI system health in real time. These indicators help pinpoint areas of success and those needing attention, allowing for timely adjustments.

Feedback Loops

Feedback loops form the communication bridge between AI systems and their human counterparts. They enable the collection and analysis of valuable insights from various sources, including direct end user feedback, system interaction data, and performance analytics. By establishing robust mechanisms to gather and interpret this feedback, organizations can uncover underlying issues, user pain points, and opportunities for system enhancement. Effective feedback loops ensure AI systems are not static entities but dynamic tools that grow and improve based on real-world usage and data.

Tools and Frameworks for Effective Monitoring and Feedback

1. **Flowcharts:**
   - Purpose: Visualize processes such as the integration steps of Generative AI or the continuous monitoring workflow.
   - Example: A flowchart mapping out each step from initial planning to real-time monitoring and feedback loops:

2. **Gantt Charts:**
   - Purpose: Schedule and track the progress of AI implementation projects over time.
   - Example: A Gantt chart illustrating timelines for each phase of AI integration, training programs, and continuous improvement cycles.

3. **Mind Maps:**
   - Purpose: Explore and visualize the relationship between different components of AI integration.
   - Example: A mind map to brainstorm and structure thoughts on cultural shifts, training needs, and stakeholder communication.

4. **Dashboard Metrics:**
   - Purpose: Provide real-time visualization of KPIs and metrics.
   - Example: A dashboard offering a comprehensive view of AI system performance, showing key metrics like processing accuracy, response times, and user satisfaction rates.

CHAPTER 8 ROADMAP FOR AI IMPLEMENTATION IN BFSI

The process of iterative improvement through feedback is where the true potential of AI in the BFSI sector is unlocked. Feedback loops provide a structured pathway for identifying and implementing necessary changes, allowing AI systems to evolve in response to new information, user needs, and technological advancements. For instance, case studies of BFSI organizations show how feedback-driven modifications have led to more intuitive user interfaces, improved risk assessment algorithms, and more personalized customer service solutions. These examples underscore the transformative impact of feedback on refining AI applications, making them more user-centric and aligned with business objectives.

Integrating continuous monitoring and feedback into the AI development cycles of the Banking, Financial Services, and Insurance (BFSI) sector is essential for the sustained relevance and efficiency of AI applications. Best practices for embedding these mechanisms include establishing clear protocols for regular performance reviews, setting up automated monitoring tools, and creating channels for feedback collection from all stakeholders. It is crucial to incorporate these practices from the initial stages of AI development and deployment, ensuring that monitoring and feedback are integral parts of the process rather than afterthoughts.

Moreover, the role of cross-functional teams cannot be overstated in this context. These teams bring together diverse perspectives, enabling a holistic interpretation of feedback and more effective implementation of changes. Their varied expertise ensures that adjustments to AI systems are technically sound, user-focused, and aligned with organizational goals.

However, establishing robust continuous monitoring and feedback systems is not without its challenges. BFSI organizations often encounter obstacles such as data silos, lack of technical expertise, and resistance to change, which can hinder the effectiveness of these systems. Additionally, ensuring the privacy and security of feedback data, especially in the context of strict regulatory environments, poses another significant challenge. To overcome these hurdles, solutions and technologies like advanced data analytics platforms, AI-powered monitoring tools, and secure feedback collection methods can be employed. These tools facilitate the gathering and analysis of vast amounts of data while streamlining the feedback process, making it more efficient and secure.

8.6. Skill Development and Hiring

Addressing the Talent Gap and Building an AI-Driven Culture in the BFSI Sector
The BFSI sector's adoption of Large Language Models (LLMs) like ChatGPT is revolutionizing operations, yet the talent gap presents a significant challenge. The specialized nature of AI technologies (Table 8-2) demands a workforce skilled in both AI principles and their applications within BFSI. Addressing this gap involves upskilling current employees and strategically hiring new talent with specialized AI expertise.

Developing AI Competencies and Reinforcing Core BFSI Skills Balancing the development of AI skills with reinforcing core BFSI competencies is crucial. Employees must become proficient in new technologies while maintaining a deep understanding of sector-specific challenges and opportunities. This dual focus ensures a workforce that is both technically adept and knowledgeable about BFSI nuances, enhancing the effectiveness of AI integration.

Building an AI Learning Culture Creating a continuous learning environment is essential for leveraging LLMs effectively. Online learning platforms and resources provide employees with access to extensive AI knowledge, from foundational concepts to advanced techniques. Flexible learning schedules facilitate balancing professional duties with personal development. Cross-functional collaboration and knowledge sharing break down silos, fostering innovation and collective problem-solving. This culture enhances AI capabilities across the organization, building a supportive community of learners.

Hiring New Talent with Specialized AI Skills Defining the ideal AI skillset for BFSI involves expertise in data analytics, machine learning, and ethical considerations unique to financial services. Effective recruitment strategies include partnerships with academic institutions, participation in tech forums, and leveraging professional networks. Integrating new talent involves aligning them with organizational goals and fostering environments where experienced professionals and AI newcomers learn from each other. This ensures new skills translate into tangible benefits, driving AI initiatives and reinforcing sector competitiveness.

Table 8-2. Key Skills

Skillset Categories	Specific Skills	Description
Technical Skills	Machine Learning (ML) and Deep Learning (DL)	Proficiency in ML and DL frameworks like TensorFlow, PyTorch, and Keras. Experience with neural networks, particularly in training and tuning LLMs.
	Natural Language Processing (NLP)	Expertise in NLP techniques and tools for text processing, sentiment analysis, and language generation. Familiarity with transformer architectures.
	Data Science and Analytics	Strong background in statistics, data analysis, and data visualization. Proficiency in Python, R, and SQL for data manipulation and model development.
	Software Development	Skills in software engineering principles, including version control (e.g., Git), testing, and deployment. Experience with APIs and cloud platforms.
	Data Management and Engineering	Knowledge of data warehousing, ETL processes, and big data technologies like Hadoop and Spark. Understanding of data privacy laws and data security best practices.
Domain-Specific Skills	Financial Knowledge	Understanding of BFSI concepts such as credit risk assessment, fraud detection, and compliance. Experience with financial data and regulatory requirements.
	Ethical AI and Bias Mitigation	Awareness of ethical considerations in AI, including bias detection and mitigation strategies. Ability to design fair and transparent AI systems.
Tools and Technologies	Frameworks	TensorFlow, PyTorch, Keras
	Languages	Python, R, SQL
	Tools	Hadoop, Spark, Git
	Cloud Platforms	AWS, Azure, Google Cloud
	NLP Libraries	spaCy, NLTK, Hugging Face Transformers

Collaboration with Academia and Industry Partnerships with universities and research institutions provide access to cutting-edge AI research and emerging talent. Collaborations can include joint research projects, internships, and specialized AI courses tailored to BFSI needs. Engaging with AI and FinTech startups brings fresh perspectives and agile solutions. Industry consortia dedicated to AI talent development offer platforms for knowledge exchange, setting industry standards, and collaborative training initiatives, enhancing overall AI expertise within the sector.

By combining these strategies—upskilling employees, hiring specialized talent, fostering an AI learning culture, and collaborating with academia and industry—BFSI organizations can effectively address the talent gap. This comprehensive approach ensures the sector fully harnesses the potential of LLMs, maintaining a competitive edge and driving sustained innovation.

Challenges and Solutions in AI Skill Development

Challenges in AI skill development within the BFSI sector include training and hiring, ensuring diversity and inclusivity, and keeping pace with technological advancements. Overcoming training and hiring obstacles requires innovative approaches, such as using online platforms for scalable learning and adopting flexible hiring practices that recognize diverse skill sets and backgrounds. Diversity and inclusivity are critical for fostering a workforce that reflects varied perspectives, enhancing AI solutions' creativity and effectiveness. This can be achieved through targeted outreach, scholarship programs, and partnerships with organizations focused on underrepresented groups in tech.

Aligning skill development with rapid technological changes necessitates a commitment to continuous learning and adaptability. Organizations must ensure that employees' skills remain relevant and can swiftly implement emerging AI technologies. This involves creating a culture that values ongoing education and adapting training programs to reflect the latest advancements.

Evaluating the Effectiveness of Skill Development Programs

Evaluating the effectiveness of AI skill development programs is crucial for ensuring that training investments deliver tangible benefits. Metrics and Key Performance Indicators (KPIs) such as employee retention rates, program completion rates, and the application of new skills in projects provide insights into the success of training initiatives. Continuous improvement approaches, including feedback loops and iterative program design, allow organizations to refine training programs based on real-world outcomes and evolving needs. Assessing the Return on Investment (ROI) involves measuring direct impacts on productivity and innovation and considering broader benefits, such as creating a technologically fluent workforce capable of driving AI-driven transformations.

8.7. Ethical and Regulatory Compliance in AI for BFSI

In the BFSI sector, integrating Generative AI presents unique ethical and regulatory challenges that organizations must carefully navigate. Leveraging AI for enhanced decision-making, streamlined operations, and improved customer experiences involves addressing issues related to bias, transparency, and accountability. Moreover, the regulatory landscape for AI in BFSI is continuously evolving, with both existing laws and emerging regulations shaping responsible AI deployment. Balancing innovation with ethical conduct and compliance is crucial to ensuring AI technologies revolutionize the sector while maintaining fairness, transparency, and privacy standards.

One of the primary ethical concerns is the risk of algorithmic biases in AI systems. These biases can skew decision-making processes, inadvertently favoring certain groups over others and compromising the fairness of financial services. To mitigate such biases, BFSI organizations should use diverse datasets for training and apply fairness algorithms to ensure AI models operate equitably. Decision transparency and explainability are also pivotal for maintaining trust. AI-driven decisions, especially those related to creditworthiness or risk assessments, must be interpretable to ensure they are justifiable. Techniques to enhance transparency include developing explainable AI models and tools that trace decision-making pathways and providing clear insights into AI-generated conclusions.

Navigating the regulatory landscape is another layer of complexity. Compliance with regulations like the GDPR requires that AI systems protect data privacy and are transparent in their operations. BFSI organizations must stay informed about current and upcoming regulations to remain compliant. Developing comprehensive compliance frameworks that include best practices for documentation, data management, and reporting is essential. Regular audits and continuous monitoring help identify compliance gaps, enabling timely remediation. Advanced data analytics platforms and AI-powered compliance tools can facilitate these efforts, streamlining the compliance process and mitigating regulatory risks.

Ethical considerations should be integrated from the outset of AI development. Conducting ethical impact assessments at various stages—from design to deployment—helps identify and mitigate potential ethical risks. Establishing and continuously updating ethical guidelines in line with technological advancements is crucial for maintaining integrity throughout the AI lifecycle. Stakeholder engagement is also vital. By involving customers, regulators, and other stakeholders, organizations gain insights into societal expectations and regulatory requirements, fostering trust and transparency.

Looking ahead, BFSI organizations must remain adaptable and proactive as societal norms and regulatory landscapes evolve. Collaboration among BFSI organizations, technology providers, regulatory bodies, and academic institutions can help develop and refine ethical guidelines and regulatory frameworks for AI. This collective effort aids in standardizing ethical AI practices across the sector and encourages innovation within a responsible and compliant framework.

8.8. Summary

1. Thoroughly evaluate existing technological infrastructure to identify areas where AI can be effectively integrated within the BFSI sector.

2. Align AI initiatives with broader business objectives to ensure they drive significant advancements rather than just incremental improvements.

3. Establish robust data governance frameworks to manage the collection, storage, and use of data, ensuring its accuracy, completeness, and compliance with regulatory requirements.

4. Implement continuous monitoring and feedback loops to regularly assess AI system performance and make necessary adjustments for optimal operation.

5. Prioritize transparency and explainability in AI decision-making processes to build trust with customers and regulators.

6. Develop strategies to address potential biases in AI models by using diverse datasets and applying fairness algorithms.

7. Ensure that AI systems comply with strict data protection regulations like GDPR to safeguard sensitive financial information.

8. Foster a culture of continuous learning and innovation within the organization to adapt to evolving AI technologies and market demands.

9. Address skill gaps by upskilling current employees and hiring specialized talent with expertise in AI- and BFSI-specific applications.

10. Establish ethical guidelines and regulatory compliance frameworks to guide the responsible development and deployment of AI systems, ensuring fairness and accountability.

CHAPTER 9

Challenges in Mainstream Adoption

To construct an informative and comprehensive Chapter 9 on the integration of artificial intelligence (AI) into the Banking, Financial Services, and Insurance (BFSI) sector, it's crucial to delve into the multifaceted challenges and opportunities this technology presents. The chapter aims to navigate through the technical, organizational, and regulatory hurdles and the nuanced market dynamics that shape the landscape of AI adoption within the sector. This exploration is not just about highlighting the hurdles but also about showcasing the innovative strides being made in overcoming these challenges, ensuring a balanced narrative that reflects both the potential and the pitfalls of AI in BFSI.

Introduction

The dawn of AI integration within the BFSI sector marks a significant paradigm shift towards enhancing operational efficiencies, customer experiences, and decision-making processes. However, this transition is accompanied by a spectrum of technical, organizational, and regulatory challenges that necessitate a nuanced approach to AI adoption. As financial institutions strive to harness the potential of AI, understanding these complexities becomes paramount to navigating the evolving landscape of financial services. This chapter aims to dissect these challenges, offering insights into the dynamic interplay between innovation and the responsible use of AI in the BFSI sector.

9.1. Challenges of Integrating AI into the BFSI Sector

1. Data Acquisition and Quality in BFSI

In the BFSI sector, obtaining high-quality financial data presents unique challenges (Table 9-1). Public financial datasets are scarce due to stringent privacy regulations

CHAPTER 9 CHALLENGES IN MAINSTREAM ADOPTION

and proprietary data, complicating AI development and necessitating specialized preprocessing to ensure data relevance. For instance, JPMorgan Chase's efforts to enhance its AI-driven investment strategies are often hampered by limited data access, requiring advanced anonymization techniques to balance data utility with privacy protection. Regulatory frameworks, including GDPR and the Dodd-Frank Act, impose strict requirements on data handling and AI transparency. Bias in AI decision-making remains a concern, leading scholars like John Smith to advocate for ethical AI development, emphasizing fairness and transparency. AI models in BFSI must be robust and reliable. The volatile nature of financial markets demands sophisticated stress testing and adaptive design, as demonstrated by Goldman Sachs' use of AI to navigate market fluctuations during economic downturns. Simulating real-world financial scenarios for AI training is crucial, incorporating diverse macroeconomic factors to accurately reflect complex financial systems. Companies like BlackRock are exploring innovative simulation techniques for effective risk management, enabling the sector to navigate financial uncertainties. Integrating AI with legacy systems poses significant challenges. Legacy systems often create data silos, hindering AI applications. Experts like Robert Lee recommend incremental strategies for integration, focusing on interoperability and modular upgrades to minimize disruptions. AI's role in fraud detection and risk management showcases its transformative potential. For example, PayPal's AI systems detect and prevent fraudulent transactions, though balancing sensitivity and specificity remains challenging. Researchers, including Sarah Brown, work on refining models to adapt to evolving fraud tactics while maintaining accuracy. Looking ahead, emerging technologies like quantum computing and blockchain promise to revolutionize the BFSI sector. These innovations enhance security, transparency, and predictive accuracy but also introduce new ethical and technical complexities. Continuous innovation and adaptability remain essential for overcoming these challenges and seizing AI's opportunities.

CHAPTER 9　CHALLENGES IN MAINSTREAM ADOPTION

Table 9-1. *Challenges in BFSI Industry*

Aspect	Challenges	Examples	Solutions/ Approaches	Technologies	Regulatory Frameworks
Data Quality	Obtaining high-quality financial data due to stringent privacy regulations and proprietary data. Bias in AI decision-making. Integrating AI with legacy systems.	JPMorgan Chase's AI-driven investment strategies, Goldman Sachs' use of AI to navigate market fluctuations, and PayPal's AI fraud detection systems.	Advanced anonymization techniques, specialized preprocessing, simulating real-world scenarios, incremental integration strategies, balancing sensitivity and specificity.	AI, advanced anonymization techniques, quantum computing, blockchain.	GDPR, Dodd-Frank Act.
Model Drift in Financial Contexts	Predictive accuracy declines as data patterns change. Continuous model maintenance is needed.	JPMorgan Chase's AI-driven investment strategies affected by evolving market conditions; Goldman Sachs' tools for real-time drift detection; PayPal's AI fraud detection systems.	Continuous monitoring and updating, automated tools, periodic retraining, feedback loops, MLOps, adherence to best practices.	MLOps, machine learning, DevOps.	Basel III.

269

Model Drift in Financial Contexts

In the BFSI sector, AI integration has transformed operations, offering advanced analytics and automated decision-making. However, financial markets' dynamic nature leads to model drift, where predictive accuracy declines as data patterns change. For example, JPMorgan Chase faces challenges maintaining AI-driven investment strategies due to evolving market conditions. This drift affects risk assessment, fraud detection, and decision-making, highlighting the need for continuous model maintenance. Detecting drift involves monitoring performance metrics like accuracy and error rates. Tools such as those used by Goldman Sachs help in real time drift detection, enabling timely model adjustments.

Continuous monitoring and updating are essential to combat model drift. Automated tools and periodic retraining ensure models stay relevant, reflecting the latest data trends. Feedback loops, where outcomes inform future predictions, help maintain accuracy. Adhering to best practices, including regular retraining and robust version control, is critical. PayPal's AI fraud detection systems exemplify the need for balance in sensitivity and specificity to adapt to new fraud patterns while avoiding false positives.

Emerging technologies like MLOps streamline AI model maintenance, combining machine learning and DevOps for efficient lifecycle management. These innovations help financial institutions like BlackRock automate model updates, ensuring compliance and resilience. Regulatory standards, such as Basel III, influence model development and maintenance, promoting transparency and consumer protection. Despite challenges, including balancing model stability and adaptability, continuous innovation and adherence to best practices are vital for leveraging AI's full potential in the BFSI sector, ensuring accuracy and compliance amidst evolving financial landscapes.

Example: Global Bank Inc. and the Challenge of Credit Risk Model Drift

Background: Global Bank Inc., a leading player in the BFSI sector, leveraged advanced AI models for credit risk assessment to determine the likelihood of borrowers defaulting on loans. These models were integral to the bank's loan approval processes, directly impacting its risk exposure and financial health.

Challenge: Over time, Global Bank Inc. observed a gradual but consistent decline in the predictive accuracy of its credit risk models. The models began to approve higher-risk loans, resulting in an uptick in default rates. An initial analysis indicated that model drift, particularly concept drift, was occurring due to changes in economic conditions, customer behavior, and market dynamics that the models hadn't been retrained to recognize.

Solution: Global Bank Inc. implemented a comprehensive strategy to address model drift, comprising several key components:

1. **Continuous Monitoring:** The bank deployed real-time monitoring tools to detect signs of model drift, using performance metrics like accuracy, precision, and recall. This allowed for the early identification of issues before they significantly impacted loan approval processes.

2. **Periodic Model Retraining:** Recognizing that static models would continually degrade in performance, Global Bank Inc. established a routine schedule for model retraining, incorporating the latest customer data and market conditions to maintain model relevance and accuracy.

3. **Version Control and Documentation:** To manage updates and maintain transparency, the bank implemented robust version control systems and comprehensive documentation practices for all AI models. This ensured traceability and facilitated regulatory compliance.

4. **Feedback Loops:** The incorporation of feedback loops allowed the bank to learn from approved and defaulted loans continuously, further refining the models with new insights and data points.

5. **Regulatory Compliance:** Adhering to regulations like Basel III, the bank ensured that its model maintenance practices were compliant, incorporating required stress testing and scenario analysis in the retraining process.

Outcome: Through these measures, Global Bank Inc. successfully managed the issue of model drift, restoring the accuracy and reliability of its credit risk models. The bank not only reduced its default rates but also improved its compliance posture, demonstrating a commitment to both innovation and regulatory adherence.

Lessons Learned: This case study underscores the importance of proactive model maintenance and the need for financial institutions to adapt to changing environments. Key takeaways for the BFSI sector include the criticality of continuous monitoring, the benefits of periodic retraining, and the value of a structured approach to version control and documentation. Moreover, it highlights the role of regulatory compliance as a driver for best practices in AI model maintenance.

CHAPTER 9   CHALLENGES IN MAINSTREAM ADOPTION

9.2. Organizational Challenges

Resistance to AI Adoption in the Banking, Financial Services, and Insurance (BFSI) Sector

In the rapidly evolving landscape of the Banking, Financial Services, and Insurance (BFSI) sector, the adoption of artificial intelligence (AI) presents a transformative opportunity for organizations seeking to enhance efficiency, customer service, and competitive edge. However, this journey is fraught with challenges, rooted in cultural, technological, economic, regulatory, and managerial dimensions. These obstacles necessitate a nuanced understanding and strategic approach to facilitate a smooth transition and maximize the benefits of AI technologies. From addressing the cultural resistance within organizations and navigating the perceived risks and concerns to overcoming technological barriers and economic considerations, BFSI institutions face a complex maze. Furthermore, navigating the regulatory and compliance uncertainties adds another layer of complexity to AI adoption. As such, employing robust change management models and frameworks, including Lewin's three-stage model, Kotter's 8-Step Process, and McKinsey's 7-S Framework, provides invaluable insights into creating a conducive environment for embracing AI, underscoring the importance of strategic alignment and adaptability in this digital age.

1. **Cultural Factors Influencing Resistance to AI in BFSI**

   In the BFSI sector, the cultural landscape plays a pivotal role in shaping attitudes towards AI adoption. Much like the sturdy oaks that stand firm against the changing seasons, traditional mindsets within these organizations often resist the winds of change brought about by AI. This resistance is not just about reluctance; it's rooted in a deep-seated fear that AI might upend established roles and practices that have been the backbone of these institutions for decades. The notion of AI, with its promise of automation and efficiency, is perceived as a threat to the human touch that has long defined customer service and decision-making in BFSI. It's akin to introducing a new species into a well-balanced ecosystem, where the existing inhabitants are wary of the newcomer's impact. Overcoming this cultural inertia requires not only demonstrating the tangible benefits of AI but also engaging in open dialogues that address fears and build a bridge between the old and the new.

CHAPTER 9   CHALLENGES IN MAINSTREAM ADOPTION

2. **Perceived Risks and Concerns**

   The path to embracing AI in the BFSI sector is fraught with perceived risks and concerns that can cloud the vision of the most forward-thinking leaders. Data security emerges as a paramount concern, akin to safeguarding a fortress in an age where digital breaches are ever more cunning and frequent. Ethical considerations, too, loom large; the prospect of AI making decisions that affect people's financial lives raises questions about fairness and bias, much like a judge tasked with making impartial decisions under the watchful eyes of society. Furthermore, the specter of job displacement creates a palpable tension, as employees fear being rendered obsolete by machines—a scenario reminiscent of the industrial revolution's impact on manual labor. Concerns about transparency and accountability in AI-driven decisions add another layer of complexity, mirroring the need for clarity and trust in the delicate dance of financial services. Addressing these concerns requires a nuanced approach that balances innovation with empathy and ethical stewardship.

3. **Technological Challenges and Complexity**

   The technological hurdles to AI adoption in BFSI are akin to navigating a labyrinth, where each turn reveals new challenges. Integrating AI with legacy systems is a daunting task, much like trying to blend modern architecture with ancient structures. It demands not only technical finesse but also a deep understanding of the old and new elements' intrinsic qualities. The need for specialized skills and knowledge adds to the complexity, highlighting a gap akin to a linguist trying to decipher an ancient language with a modern dictionary. Furthermore, the sheer complexity of managing advanced AI technologies can feel like holding a map without a compass; the path is charted, but the direction is unclear. Bridging these gaps requires concerted efforts to demystify AI, cultivate technical expertise, and ensure a seamless melding of technologies.

4. **Economic Considerations and Cost Implications**

   The economic landscape surrounding AI adoption in BFSI is dotted with concerns about costs, Return on Investment (ROI), and financial risks, reminiscent of a ship navigating through treacherous waters. The initial outlay for implementing AI technologies can be steep, evoking fears akin to placing a bet on an untested horse. Doubts about the ROI further compound this hesitancy, as organizations grapple with the uncertainty of whether the investment will bear fruit or turn into a financial sinkhole. The risks associated with technology adoption, from unforeseen costs to potential failures, add another dimension to the economic deliberations, akin to a farmer weighing the risk of planting a new crop that could either flourish or falter. Navigating these economic considerations demands a balanced approach that weighs immediate costs against long-term value, ensuring that the journey towards AI is both financially viable and strategically sound.

5. **Regulatory and Compliance Issues**

   The regulatory and compliance landscape in BFSI is a moving target, complicating the adoption of AI. Concerns about meeting ever-evolving regulatory standards are like navigating a maze that constantly changes its layout. Institutions must juggle the dual challenge of harnessing AI's potential while ensuring that these technologies do not run afoul of regulations, a balancing act akin to walking a tightrope with regulatory watchdogs on either side. The potential legal implications of AI decisions introduce a layer of complexity that echoes the cautionary tales of early explorers charting unknown territories, where each step could lead to unforeseen consequences. Addressing these regulatory challenges requires a proactive stance, where institutions not only comply with current regulations but also anticipate future shifts in the regulatory landscape, ensuring a smooth sail in the dynamic seas of BFSI.

Case Study 1

To navigate the complex process of AI adoption in the BFSI sector, we can apply Kurt Lewin's three-stage model of change management, which includes unfreezing, changing, and refreezing. Let's illustrate this with an example involving "Global Bank," a traditional banking institution grappling with the transition towards AI.

Unfreezing

In this initial stage, Global Bank recognizes the need to adopt AI to stay competitive and meet changing customer expectations. However, there's a thick layer of resistance, primarily due to a traditional mindset and fear among employees about job security. To unfreeze this inertia, the bank's leadership begins a comprehensive communication strategy. They organize workshops and seminars, akin to preparing the soil before planting seeds, where the focus is on showcasing the benefits of AI, such as improved efficiency and enhanced customer service. Simultaneously, they address fears directly by highlighting how AI will augment jobs rather than replace them, using analogies like the introduction of ATMs, which changed the role of bank tellers but didn't make them obsolete.

Changing

Once the ground is fertile with openness to change, Global Bank starts implementing AI in phases, focusing initially on back-office operations to streamline processes and reduce errors. This phase is akin to planting the seeds of change and nurturing them with careful attention. The bank invests in training programs to equip employees with the necessary skills to work alongside AI systems, emphasizing the symbiotic relationship between human creativity and AI's analytical prowess. They also initiate pilot projects in select branches, allowing employees to experience firsthand how AI can enhance their work, not unlike chefs experimenting with new ingredients to enhance traditional recipes. During this phase, feedback mechanisms are crucial, serving as the water that helps the seeds of change grow. Employees are encouraged to share their experiences and concerns, ensuring that the process is collaborative and inclusive.

Refreezing

As the new AI systems begin to show positive results, Global Bank moves towards the refreezing stage, where these changes are solidified into the organization's new normal. This stage is akin to the seeds finally taking root and growing into sturdy plants. The successful pilot projects are scaled up, and AI integration becomes part of the bank's standard operating procedures. To ensure the changes stick, Global Bank updates its policies and training programs to reflect the new AI-enhanced workflows. Recognition programs are introduced to celebrate the successes and contributions of teams and individuals in embracing AI, much like gardeners admiring the fruits of their labor. This not only cements the new practices but also reinforces the culture of innovation and adaptability within the organization.

CHAPTER 9   CHALLENGES IN MAINSTREAM ADOPTION

Throughout this transition, Global Bank faces counterarguments, such as skepticism about AI's reliability and ethical concerns. By engaging these viewpoints openly, conducting thorough risk assessments, and ensuring transparent decision-making processes, the bank navigates these challenges thoughtfully, ensuring that the shift towards AI is both effective and ethically sound.

Case Study 2

Let's explore Kotter's 8-step process for leading change through the lens of a BFSI institution, "FutureBank," as it embarks on a transformative journey to integrate AI across its operations.

1. **Creating a Sense of Urgency**

    FutureBank's leadership, led by CEO Emily Chen, begins by highlighting the rapid evolution of the financial sector due to technological advancements. In town hall meetings and communications, they paint a vivid picture of the competitive landscape, using market trends and consumer expectations as a backdrop. This approach, akin to sounding an alarm ahead of a storm, aims to awaken the organization to the critical need for change.

2. **Building a Guiding Coalition**

    Recognizing the importance of collective effort, Emily forms a diverse team of change champions from various departments, including tech-savvy innovators and respected traditionalists like CFO Michael Johnson. This coalition serves as the backbone for the initiative, embodying the fusion of old and new, much like an orchestra that blends traditional and modern instruments to create a harmonious symphony.

3. **Forming a Strategic Vision and Initiatives**

    With the coalition's input, FutureBank crafts a vision that articulates how AI can enhance customer service, streamline operations, and foster innovation. They outline specific initiatives, such as implementing AI in risk assessment and personalized banking services, illustrating these goals with scenarios that resonate with employees' daily experiences, making the abstract concept of AI as tangible as the tools they use in their work.

4. **Enlisting a Volunteer Army**

   To drive grassroots support, the coalition encourages enthusiastic employees to take active roles in AI projects. This "volunteer army" becomes pivotal in spreading excitement and ownership of the change process, much like community leaders who inspire their neighbors to participate in a local initiative, fostering a sense of belonging and contribution.

5. **Enabling Action by Removing Barriers**

   FutureBank undertakes an audit to identify and dismantle obstacles to AI integration, ranging from outdated processes to skepticism. By addressing these barriers directly—such as offering reskilling programs and creating platforms for open dialogue—they clear the path for change, akin to removing weeds from a garden to allow new plants to flourish.

6. **Generating Short-Term Wins**

   The bank launches pilot AI projects in select branches and departments, achieving early successes in areas like fraud detection and customer service personalization. These victories are widely celebrated, serving as proof of concept and building momentum, much like a series of small victories in a larger campaign that boosts morale and confidence.

7. **Sustaining Acceleration**

   Buoyed by these initial wins, FutureBank accelerates the rollout of AI, expanding projects while continuously seeking feedback and adapting strategies. This phase is reminiscent of a snowball rolling downhill, gathering size and speed, as the organization harnesses the energy of early successes to propel further innovation.

8. **Instituting Change**

    Finally, the successful integration of AI becomes a part of FutureBank's DNA, reflected in updated policies, training programs, and the incorporation of AI-driven decision-making into strategic planning. This institutionalization ensures that the change is not a fleeting initiative but a lasting part of the bank's identity, much like a town adopting new traditions that enrich its culture.

Throughout this journey, FutureBank navigates skepticism and resistance by engaging critics in constructive dialogues, ensuring transparency in decision-making, and demonstrating the tangible benefits of AI. By meticulously following Kotter's 8-step process, FutureBank transforms the challenge of AI integration into an opportunity for growth and innovation, setting a new standard for the BFSI sector.

McKinsey 7-S Framework

Let's apply the McKinsey 7-S Framework to a fictional example involving "TechNova Bank" (Table 9-2), a BFSI organization looking to integrate AI across its operations to enhance efficiency, improve customer service, and stay competitive.

Table 9-2. McKinsey 7-S Framework

Aspect	Description
Strategy	Leveraging AI for risk management, customer personalization, and operational efficiency.
Structure	Introducing an AI innovation department headed by Mark Johnson for cross-functional collaboration.
Systems	Upgrading IT infrastructure and performance management systems to support AI initiatives.
Shared Values	Commitment to innovation and customer service excellence, viewing AI as a tool to enhance human elements.
Skills	Comprehensive training programs for AI basics and advanced analytics, equipping employees with necessary skills.
Style	Participative and transparent leadership by Laura Smith, fostering trust and inclusivity.
Staff	Reevaluating staffing needs, addressing job displacement concerns, and creating new positions focused on AI development and ethics, offering reskilling opportunities.

CHAPTER 9   CHALLENGES IN MAINSTREAM ADOPTION

Strategy

TechNova Bank, led by CEO Laura Smith, devises a clear strategy focused on leveraging AI for risk management, customer personalization, and operational efficiency. This strategy, akin to plotting a course through uncharted waters, aims to ensure that every initiative aligns with the bank's goal of becoming a leader in AI-driven banking services.

Structure

To support this strategy, the bank reevaluates its organizational structure. It introduces a new AI innovation department, headed by a seasoned leader, Mark Johnson, facilitating cross-functional collaboration. This structural adjustment is much like renovating a building to make it more suitable for modern needs, ensuring that the flow of information and decision-making supports AI integration.

Systems

TechNova Bank updates its systems to support AI initiatives, from IT infrastructure to performance management systems. This involves not just technical upgrades but also the introduction of processes that encourage innovation and agile response to challenges, similar to installing a new operating system on a computer to improve performance and capability.

Shared Values

Central to the bank's ethos is a commitment to innovation and customer service excellence. Laura and her team work tirelessly to ensure these shared values permeate every level of the organization, reinforcing the belief that AI is a tool to enhance, not replace, the human elements of banking. This effort is akin to nurturing a garden, where the values are the sunlight and water that help the organization grow in the desired direction.

Skills

Recognizing the importance of having a workforce capable of implementing and working alongside AI, TechNova Bank launches comprehensive training programs. These are designed to upskill employees, from AI basics for all staff to advanced analytics for specialized teams. It's similar to training athletes for a new sport, equipping them with the skills needed to excel.

Style

Laura's leadership style is participative and transparent, encouraging open dialogue about AI's benefits and challenges. This approach fosters a culture of trust and inclusivity, much like a coach who nurtures a team spirit, making every team member feel valued and part of the AI journey.

Staff

The bank also reevaluates its staffing needs, identifying roles that AI can enhance and creating new positions focused on AI development and ethics. This process is handled with sensitivity to current employees' concerns about job displacement, offering reskilling opportunities, and clear communication about the bank's future direction.

In navigating the adoption of AI, TechNova Bank faces its share of resistance, from skepticism about AI's reliability to fears of job loss. However, by ensuring alignment among the 7-S elements, the bank creates a coherent and supportive environment for change. Each element, from strategy and structure to shared values and staff, is like a thread in a tapestry, woven together to create a resilient and adaptable organization ready to embrace the possibilities of AI. Through this holistic approach, TechNova Bank not only reduces resistance but also fosters a culture of innovation and continuous learning.

9.3. Training Needs for AI Integration in the Banking

The integration of artificial intelligence (AI) in the Banking, Financial Services, and Insurance (BFSI) sector marks a significant evolution in how financial services operate, innovate, and serve their customers. However, the seamless adoption and effective utilization of AI technologies hinge on addressing the pressing challenge of skill and knowledge gaps within the workforce. Identifying these gaps is the first crucial step towards crafting a comprehensive strategy for workforce development, encompassing both technical and soft skills that are pivotal for navigating the AI landscape. Employee training emerges as a linchpin in this strategy, offering a direct path to enhancing engagement, reducing resistance, and elevating the efficacy of AI implementations. The breadth of training needs spans from foundational awareness for non-technical staff to advanced technical competencies for specialized teams. Adopting a role-based training approach ensures that these initiatives are precisely aligned with the diverse roles and responsibilities across the BFSI sector. Moreover, leveraging existing frameworks for AI skill development, embracing continuous learning paradigms, forging partnerships with academic and technological entities, and deploying robust measures to gauge training effectiveness collectively form a holistic blueprint for empowering the BFSI workforce for the AI era.

1. **Identification of Skill and Knowledge Gaps**

   A thorough assessment reveals significant gaps in essential technical skills such as data analytics, machine learning, and coding. Moreover, there is an evident need for soft skills, including critical thinking and adaptability. These gaps are akin to missing pieces in a puzzle that, once filled, can create a complete picture of effective AI integration. For instance, without the ability to interpret complex data sets or adapt to new technologies, employees may struggle to utilize AI tools effectively, much like a sailor navigating without a compass.

2. **Importance of Employee Training in AI**

   The role of employee training in this context cannot be overstated. Just as watering and sunlight are vital for the growth of a plant, training is crucial for nurturing the workforce's ability to adapt and thrive alongside AI. Training enhances employee engagement, acting as a bridge over the chasm of resistance to technology. It transforms apprehension into action, equipping the workforce with the knowledge and skills to harness AI solutions effectively.

3. **Training for Non-technical Staff**

   For non-technical staff, the focus of training shifts towards building a foundational understanding of AI capabilities and its applications within their specific roles. Basic data literacy becomes a key component, ensuring that all employees, regardless of their technical background, can appreciate the value and potential of AI. This approach democratizes AI knowledge, much like teaching basic health practices improves overall community well-being.

4. **Advanced Training for Technical Teams**

   Conversely, the training needs of technical teams delve into the deep end of the AI pool. Specialized skills in AI and machine learning, along with data science competencies, become the focus. Integrating AI with existing BFSI technologies requires a nuanced understanding of both the new and old systems, reminiscent of an architect designing modern additions to a historic building. This advanced training ensures that the technical teams are not just participants in the AI journey but are leading the charge.

5. **Role-Based Training Approach**

    Adopting a role-based training approach in the BFSI sector is akin to crafting a bespoke suit; it ensures the training curriculum precisely fits the specific roles and responsibilities of employees. This method acknowledges that the needs of a data scientist are vastly different from those of a customer service representative. Just as a tailor adjusts every stitch to suit the wearer, role-based training tailors content to address the unique requirements of each position, enhancing relevance and application of AI skills in daily tasks. This approach maximizes training efficiency and effectiveness, ensuring employees not only understand AI but can also leverage it in ways that directly enhance their performance and contribution to the organization.

6. **Existing Frameworks for AI Skill Development**

    Several frameworks and models have emerged to guide AI skill development in BFSI, serving as blueprints for constructing a competent AI-ready workforce. Industry-specific competency frameworks outline the essential skills and knowledge bases required across different roles, acting as roadmaps for employee development. Professional development models provide a structured pathway for continuous skill enhancement, while standardized AI certification programs offer benchmarks for skill validation. These frameworks, much like the guidelines for constructing a sturdy building, ensure that the development of AI capabilities is systematic, comprehensive, and aligned with industry standards.

7. **Incorporating Continuous Learning and Development**

    In the fast-paced world of AI, continuous learning and development are non-negotiable for staying ahead. Strategies such as microlearning—bite-sized, easily digestible learning modules—on-the-job training, and regular workshops ensure that learning becomes an integral part of the daily routine, much like a gardener regularly tending to plants. This approach keeps the workforce agile and responsive to new developments, ensuring that BFSI organizations remain competitive and innovative. Continuous learning cultivates a culture of curiosity and adaptability, essential traits for navigating the AI landscape.

8. **Partnerships with Educational Institutions and Tech Firms**

    Forging partnerships with educational institutions and technology firms opens a conduit for specialized AI training and upskilling programs. These collaborations bring fresh perspectives and cutting-edge knowledge directly into BFSI organizations, akin to cross-pollination in nature that enhances biodiversity. By leveraging the expertise and resources of academic and tech leaders, BFSI organizations can offer their employees access to state-of-the-art training, ensuring that their skills are not only current but also aligned with the latest industry trends and technologies.

9. **Measuring Training Effectiveness**

    Evaluating the effectiveness of AI training programs is critical to ensuring that the investment in employee development yields tangible benefits. Methods for assessment might include measuring skill acquisition through tests and practical assignments, analyzing changes in productivity, and observing the impact on AI adoption and integration within the organization. These metrics serve as the compass that guides training programs, ensuring they are directionally correct and aligned with organizational goals. Just as a ship's captain adjusts course based on navigation data, BFSI organizations must use these metrics to refine and optimize their training initiatives, ensuring that their workforce is not only prepared for the present but also equipped for the future.

CHAPTER 9   CHALLENGES IN MAINSTREAM ADOPTION

10. **Market and Customer-Related Challenges**

In the ever-evolving landscape of the BFSI sector, AI reshapes customer interaction, service delivery, and trust dynamics. Trust, the bedrock of finance, faces new challenges and opportunities with AI (Table 9-3). Generative AI and Large Language Models (LLMs) can enhance customer trust through improved efficiency, personalized services, and advanced security, but they can also undermine trust if transparency and privacy concerns aren't addressed. Understanding AI's impact on trust is crucial for BFSI institutions. This exploration delves into perceptions of AI services, the role of transparency, privacy concerns, and the balance between personalization and privacy.

Table 9-3. *Market and Customer-Related Challenges*

Aspect	Advantages	Challenges
Service Efficiency	24/7 customer service, instant responses, streamlined transactions.	Potential inaccuracies, reliance on technology.
Personalization	Tailored financial advice, personalized investment opportunities.	Possible intrusion into personal data, perceived privacy issues.
Security	Advanced fraud detection, enhanced security measures.	Risk of data breaches, cybersecurity threats.
Transparency	Explainable AI providing clear decision-making processes.	Lack of understanding of AI decisions, opaque algorithms.
Data Privacy	Better control over data use, personalized yet secure services.	Fear of data misuse, concerns over identity theft.
Human Interaction	Instant support through chatbots and virtual assistants.	Lack of personal touch, frustration with automated responses.
Trust	Accurate risk assessments, reliable financial advice.	Erosion of trust due to AI malfunctions, incorrect information.

Generative AI and LLMs significantly enhance customer trust through improved service efficiency, personalized offerings, and advanced security measures. These AI systems can provide instant, accurate responses to customer inquiries, offer personalized financial advice, and detect fraudulent activities with greater precision. This can lead to increased customer satisfaction and confidence in financial institutions. For instance, timely, relevant investment advice from an AI system can boost customer confidence. However, if these technologies lack transparency or malfunction, they can erode trust. If a chatbot provides incorrect information, customers may doubt the system's reliability.

Customer perceptions of AI-driven services vary widely, influenced by media portrayal, personal experiences, and societal trends. Sensational media reports about AI might paint a futuristic world where human roles are minimized, sparking fear and mistrust. However, positive personal experiences with AI, such as the convenience of voice-activated assistants for simple banking queries, can significantly shift this perception. Familiarity is key; the more customers interact with AI and see its benefits, the more their trust tends to grow.

Transparency in AI systems is essential for trust. Explainable AI, which makes AI decisions understandable, is crucial. In loan applications, for instance, explainable AI provides clear reasons for decisions, maintaining trust in fairness and accuracy. Customers may become wary if they do not understand how AI systems make decisions or if they feel their personal data is being mishandled. Transparency about data usage and opt-out options helps maintain this delicate balance.

Data privacy and security are paramount in maintaining customer trust, especially in an era where data breaches and identity theft are not uncommon. The use of personal data in AI algorithms can be a double-edged sword. Personalized investment advice based on AI's analysis of individual spending patterns can feel like a breach of privacy to some customers. Ensuring robust cybersecurity measures and transparent data usage policies is crucial. A breach, such as the infamous Equifax incident, can have a lasting impact on customer trust across the sector.

AI-driven personalization in financial services walks a tightrope between offering tailored services and respecting privacy. Customers appreciate when a banking app alerts them to potential savings opportunities based on their spending habits, demonstrating a balance between personalization and privacy. However, the key is consent; customers must feel in control of their data and understand how it's being used to tailor services. Transparency about data usage, coupled with the ability to opt-out, helps maintain this delicate balance.

Consumer behavior studies focusing on AI in the financial services sector have become increasingly important as BFSI organizations strive to understand and adapt to customers' evolving needs and attitudes. Research on customer attitudes towards robo-advisors, AI-driven customer service, and automated investment platforms sheds light on the nuanced relationship between technology adoption and customer satisfaction.

For instance, studies on robo-advisors, which offer automated, algorithm-driven financial planning services with minimal human supervision, have shown a mixed reception among consumers. While some appreciate the convenience, accessibility, and lower costs associated with robo-advisors, others express concerns over the lack of personalized advice and the ability to handle complex financial situations. This dichotomy highlights the importance of blending AI capabilities with human expertise to cater to a broader range of customer preferences and needs.

Research on AI-driven customer service, such as chatbots and virtual assistants, indicates a growing acceptance among consumers, especially for routine inquiries and transactions. Customers value the speed and efficiency provided by AI solutions, which can offer instant responses to their queries around the clock. However, these studies also underscore the critical need for these AI systems to accurately understand and respond to customer queries to prevent frustration and erosion of trust.

Automated investment platforms, another area of interest, have garnered attention for their ability to democratize investment advice and manage portfolios with precision. They attract customers with low fees and personalized management but face concerns about rapid market changes and impersonal service. These studies emphasize the need for BFSI organizations to invest in AI technologies that enhance service delivery while maintaining a human touch where needed. The following points represent the concerns from the industry:

1. Generative AI and Large Language Models (LLMs) are leading the current technological transformation in various industries.

2. These advanced AI systems have the capability to generate human-like text and understand context with remarkable accuracy.

3. Financial institutions are significantly impacted by these technologies, as they reshape operations and customer engagement.

4. The incorporation of generative AI and LLMs offers substantial opportunities for innovation and efficiency in the financial sector.

5. Alongside the benefits, these technologies also introduce challenges, particularly related to trust in AI-driven decisions.

6. Transparency in how generative AI and LLMs operate is crucial to maintaining user confidence and regulatory compliance.

7. Data privacy concerns arise with the extensive use of these AI systems, requiring careful management to protect sensitive information.

8. Balancing the opportunities and challenges of generative AI and LLMs is essential for their responsible integration into financial services.

Trust is the foundation of the financial sector, akin to the base of a towering skyscraper; without it, the entire structure is at risk. Generative AI and LLMs add a new layer to this foundation, offering both reinforcement and potential fault lines. AI can streamline transactions, provide more accurate risk assessments, and offer 24/7 customer service through sophisticated chatbots. For instance, a customer who receives timely and relevant investment advice from an AI system is likely to feel more confident in their financial decisions. However, if AI applications lack transparency or malfunction, they can sow seeds of doubt. A notable example is when customers receive incorrect information from a chatbot, leading them to question the reliability of the entire system.

Consumer behavior studies on AI-driven financial services further illuminate the importance of trust. These studies show that while AI can enhance service efficiency and personalization, it must be implemented in a way that maintains transparency and respects privacy. Customers' acceptance of AI technologies is often contingent on their understanding and control over how their data is used. For instance, customers are more likely to trust AI-driven financial advice if they know the criteria behind the recommendations and feel their data is secure.

These studies also highlight the need for BFSI institutions to invest in AI technologies that enhance service delivery and maintain a human touch. For example, robo-advisors offering automated financial planning services have shown mixed receptions among consumers. Some appreciate their convenience and low cost, while others miss personalized advice for complex situations. Blending AI with human expertise caters to diverse preferences and needs. Similarly, AI-driven customer service, such as chatbots and virtual assistants, is increasingly accepted for routine inquiries due to its speed and efficiency, but accuracy is critical to prevent frustration.

Automated investment platforms have democratized investment advice, managing portfolios with precision and scalability. They attract customers with low fees and personalized management but face concerns about rapid market changes and the impersonal nature of the service. This highlights the importance of balancing technological efficiency with personal interaction to maintain trust.

9.4. Data Security Concern

The integration of AI into the BFSI sector represents a leap in operational efficiency, customer service, and innovative solutions. However, this advancement brings complex data security concerns. AI systems processing vast amounts of financial data introduce new vulnerabilities and challenges, including increased attack surfaces and the complexity of AI algorithms. Robust cybersecurity measures tailored to AI-driven environments are essential. AI technologies introduce specific risks, such as adversarial attacks that trick AI systems into making incorrect decisions. The theft of AI models poses another threat, revealing valuable intellectual property. AI can also bypass traditional security measures by exploiting system loopholes. These vulnerabilities require updated security strategies.

Data privacy challenges emerge as AI systems collect and process extensive customer data. Protecting this data and complying with privacy regulations like GDPR and CCPA is crucial. Ensuring AI systems respect privacy and employ secure data handling practices maintains trust. Regulatory compliance for AI in BFSI is complex, with evolving frameworks requiring transparency, accountability, and ethical AI use. BFSI institutions must adapt to these changing regulations to protect consumer rights.

AI also enhances cybersecurity by offering advanced threat detection and response capabilities. Unlike traditional methods, AI analyzes patterns and identifies anomalies, detecting sophisticated attacks. It can automate responses, isolating threats faster than manual intervention.

Existing cybersecurity protocols like PCI DSS and GLBA provide a foundation but need updates to address AI-specific challenges. Emerging trends focus on securing AI algorithms, ensuring decision integrity, and safeguarding data. Ethical AI use in data security involves maintaining transparency, preventing bias, and ensuring accountability. Continuous monitoring and updates are essential for addressing new threats and ethical concerns. These perspectives highlight the dual role of AI in presenting risks and offering solutions, underscoring the need for innovative, dynamic cybersecurity measures in the BFSI sector.

9.5. Regulatory and Compliance Challenges

The integration of artificial intelligence (AI) into the Banking, Financial Services, and Insurance (BFSI) sector marks a pivotal shift towards digital innovation, promising enhanced operational efficiency, improved customer service, and novel product offerings. Recognizing the profound implications of AI, the Reserve Bank of India (RBI) has proactively outlined a comprehensive regulatory framework aimed at guiding BFSI institutions in the responsible deployment of AI technologies. This framework underscores the RBI's commitment to fostering innovation while ensuring consumer protection, data privacy, and systemic stability. The primary concerns articulated by the RBI revolve around data security, algorithmic transparency, ethical use of AI, and the mitigation of potential risks associated with AI deployment. By establishing clear objectives and principles for AI integration, the RBI aims to navigate the BFSI sector through the complexities of digital transformation, ensuring that AI technologies are leveraged in a manner that aligns with broader financial sector goals and regulatory standards.

Concerns Raised by the Reserve Bank of India (RBI) Regarding AI Integration in Financial Institutions

The Reserve Bank of India (RBI) has articulated several concerns regarding the deployment of artificial intelligence (AI) within financial institutions, spotlighting the critical need for a balanced and cautious approach to AI integration in business processes and decision-making. The concerns raised span from design-specific risks, such as biases and robustness issues, to traditional and user-specific challenges, including data privacy, cybersecurity, consumer protection, and preserving financial stability. These concerns are categorized into two main areas: data bias and robustness, and governance.

Data Bias and Robustness

The RBI emphasizes that AI's effectiveness is contingent on the quality of its training data, inheriting any biases, errors, and issues present. Unlike humans, who derive conclusions from a lifetime of experiences and are capable of collaborative problem-solving, AI models operate on a fundamentally different basis. While human decision-making incorporates checks and balances to mitigate biases, AI models, particularly those altering parameters iteratively, often become "black boxes" that challenge audit and supervisory review. The RBI highlights the necessity for financial institutions to be vigilant about various risks associated with AI models, including but not limited to arbitrary code execution, data poisoning, data drift, unexpected behavior, and biased predictions.

CHAPTER 9   CHALLENGES IN MAINSTREAM ADOPTION

Governance

AI introduces novel challenges in governance, especially as it enables autonomous decision-making that could reduce or eliminate human judgment and oversight. Issues such as prompt injection, hallucinations, and toxic outputs not only affect the models but also have significant implications for governance frameworks within financial institutions. This situation calls for a reevaluation of existing regulatory frameworks to ensure they adequately cover consumer protection, cybersecurity, and data privacy in the context of AI use.

Expectations from Financial Institutions

Given these concerns, the RBI sets forth expectations for financial institutions incorporating AI into their operations. It advocates for the implementation of robust governance frameworks that address the unique challenges posed by AI, including enhanced transparency, rigorous model validation processes, and the establishment of ethical guidelines to govern AI use. Financial institutions are expected to develop and adhere to practices that ensure fairness, prevent biases, and maintain the integrity and security of consumer data. Moreover, institutions are encouraged to foster an environment of continuous learning and adaptation to evolving AI technologies and regulatory landscapes.

RBI Guidelines on AI Implementation and Compliance (Here)

In the realm of deploying artificial intelligence (AI) models within financial institutions, the Reserve Bank of India emphasizes the importance of balancing technological innovation with the responsible use of AI. To safeguard fairness, mitigate biases, and protect consumer privacy, the following guidelines have been established:

1. **Equity and Fairness**: Financial institutions are mandated to ensure that AI algorithms do not discriminate unlawfully against individuals based on protected characteristics. Regular audits, including third-party validations, should be conducted to detect and correct any biases, thereby upholding ethical standards.

2. **Clarity and Transparency**: It is essential that all parties involved have a clear understanding of the inputs and processes leading to AI-driven decisions. Efforts must be made to demystify the decision-making process, making it accessible and comprehensible to both regulators and consumers alike.

3. **Precision and Accuracy**: Institutions must employ precise and relevant training data to reduce decision-making errors. Ongoing efforts to identify and decrease both false positives and negatives are crucial for the integrity of AI applications.

4. **Uniformity and Consistency**: The consistent application of AI algorithms is required to eliminate biases and ensure fair outcomes across diverse scenarios. Stability in the parameters used by the models is crucial, avoiding arbitrary adjustments that may compromise fairness.

5. **Privacy and Data Protection**: Adhering to stringent data protection laws and handling personal information with the highest security measures is non-negotiable. AI models must be designed to comply with existing data protection frameworks, ensuring responsible management of personal data.

6. **Explainability and Comprehension**: Financial entities must be capable of elucidating the determinants that influence AI-driven outcomes. Establishing a transparent process and providing avenues for addressing inquiries or disputes are vital in fostering trust in AI systems.

7. **Accountability for AI Decisions**: Clear accountability mechanisms must be in place, delineating responsibility for the AI model's performance, fairness, and resilience. A comprehensive governance structure, including regular evaluations and compliance checks, is required to address and rectify any issues.

8. **Durability and Resilience**: Rigorous testing to confirm the AI algorithm's performance across various conditions is essential. Financial institutions should regularly refresh training data and ensure the model's adaptability to evolving economic environments.

9. **Continuous Oversight and Adaptation**: Ongoing monitoring of AI systems is imperative to accommodate market changes and emerging challenges. Especially critical is the vigilance over self-learning algorithms to maintain their original performance and ethical standards.

10. **Incorporation of Human Judgment**: Integrating human oversight is crucial for addressing complex or nuanced cases, ensuring that ethical considerations are always at the forefront. This oversight aids in the timely identification and resolution of any unforeseen outcomes or governance issues.

As the BFSI sector continues to evolve under the influence of AI and related technologies, the RBI's regulatory framework serves as a beacon, guiding institutions through the challenges of compliance while seizing the opportunities that AI offers. The guidelines issued by the RBI emphasize the importance of robust risk management practices, adherence to ethical standards, and the safeguarding of consumer interests. By drawing comparisons with global regulatory frameworks and learning from diverse case studies, BFSI institutions can gain valuable insights into navigating the AI regulatory landscape effectively. Furthermore, the role of international collaboration in shaping future regulatory approaches highlights the global nature of the challenges and opportunities presented by AI. In conclusion, as BFSI institutions strive to align with the RBI's guidelines, they are encouraged to embrace a culture of innovation, transparency, and ethical responsibility. This approach not only ensures compliance but also positions these institutions to compete successfully on a global stage, leveraging AI to drive growth, enhance customer experiences, and contribute to the overall stability and prosperity of the financial sector.

9.6. Summary

1. Emerging technologies are driving transformative changes across industries, enhancing efficiency, personalization, and innovation.

2. Addressing ethical challenges like bias and fairness is crucial for responsible development and deployment of new technologies.

3. Adhering to regulations and guidelines ensures transparency, consumer protection, and the upholding of industry standards.

4. Strong governance frameworks and oversight mechanisms are essential for aligning technological advancements with ethical and societal values.

CHAPTER 9   CHALLENGES IN MAINSTREAM ADOPTION

5. Automation and innovation are reshaping the workforce, necessitating reskilling and upskilling to adapt to new roles.

6. Balancing rapid innovation with ethical responsibility is key to ensuring technological progress benefits all stakeholders.

7. Anticipating future trends in emerging technologies is important for ongoing transformation and industry evolution.

8. Learning from past technological failures provides valuable lessons for improving future outcomes.

9. Protecting data and privacy is increasingly critical in the digital age, requiring robust strategies to ensure security.

10. Strategic governance plays a crucial role in responsibly implementing technological innovations and fostering trust and sustainability.

CHAPTER 10

Ethical Dilemmas and Future Potential of Generative AI in the Financial Sphere

Introduction

The integration of artificial intelligence (AI) in the Banking, Financial Services, and Insurance (BFSI) sector promises transformative benefits, from enhancing operational efficiency to personalizing customer experiences. Generative AI, a rapidly evolving subset of AI, is at the forefront of this revolution, capable of creating new content and insights from vast datasets. However, the adoption of such advanced technologies is not without significant ethical and responsible considerations. This book chapter delves into the multifaceted ethical dimensions of AI in the BFSI sector, exploring the challenges, real-world consequences, responses from the tech industry, BFSI-specific nuances, and governance frameworks essential for guiding AI's future trajectory. Through detailed case studies and expert insights, this chapter emphasizes the critical need for a responsible and transparent approach to AI integration in financial contexts.

CHAPTER 10   ETHICAL DILEMMAS AND FUTURE POTENTIAL OF GENERATIVE AI IN THE FINANCIAL SPHERE

10.1. Ethical and Responsible Concerns in Generative AI

Generative AI, a rapidly advancing subset of artificial intelligence, is designed to create new content by learning from vast datasets. Unlike traditional AI, which typically analyzes existing data, generative AI can generate human-like text, images, audio, and even complex financial models. This capability has significant implications for the Banking, Financial Services, and Insurance (BFSI) sector, transforming operations, enhancing customer interactions, and driving innovation. For instance, JPMorgan Chase developed COiN (Contract Intelligence), an AI system that reviews legal documents and extracts essential data, significantly reducing processing time and operational costs. This system can analyze complex legal contracts in seconds, a task that would take humans much longer to complete. Similarly, HSBC has improved its fraud detection capabilities by leveraging AI to analyze transaction patterns and detect anomalies, thereby enhancing security and customer trust.

In addition to fraud detection and document processing, generative AI is being used to personalize customer interactions and streamline banking processes. Capital One, for example, uses AI to analyze customer data and offer personalized financial advice. The AI system tailors recommendations for credit cards, loans, and other financial products, thus enhancing customer satisfaction and engagement by making financial services more accessible and relevant to individual needs. BBVA has integrated generative AI into its loan approval process, which has reduced the time required for approvals and improved the accuracy of risk assessments. By utilizing AI to evaluate loan applications more efficiently, BBVA can better manage risk and serve its clients more promptly.

The Commonwealth Bank of Australia (CBA) employs AI-driven chatbots and virtual assistants to handle customer inquiries, significantly improving service efficiency. These AI systems can address a wide range of customer questions quickly and accurately, freeing up human agents to focus on more complex issues. This not only enhances customer service but also optimizes resource allocation within the bank. The adoption of generative AI in these various applications demonstrates its broad utility in the BFSI sector, highlighting how the technology drives innovation and operational efficiency while improving customer experiences.

Moreover, the benefits of generative AI extend beyond individual customer interactions to more strategic applications. For example, AI-driven predictive models are transforming investment strategies. Generative AI can simulate market conditions

CHAPTER 10   ETHICAL DILEMMAS AND FUTURE POTENTIAL OF GENERATIVE AI IN THE FINANCIAL SPHERE

and generate potential investment opportunities by analyzing historical market data and financial indicators. This helps investors optimize their portfolios and make informed decisions, illustrating the technology's capability to support high-level financial planning and risk management. Additionally, AI systems are being used to develop personalized investment portfolios based on an individual's financial behavior and goals, thereby improving customer satisfaction and engagement.

Ethical considerations, however, are crucial to prevent misuse and bias. Generative AI systems can inadvertently perpetuate existing biases present in the training data, leading to unfair treatment of certain demographic groups, especially in sensitive areas like credit scoring and loan approvals. For instance, if historical data reflects systemic biases, the AI might unfairly disadvantage certain demographic groups. Addressing these issues involves implementing robust bias detection and mitigation strategies to ensure fairness and equity in AI outcomes. Moreover, handling vast amounts of customer data raises significant privacy concerns. Financial institutions must manage this data responsibly, complying with data protection regulations to maintain customer trust. Transparency and explainability are also critical, ensuring that AI systems' decision-making processes are understandable to both customers and regulators.

In conclusion, generative AI holds immense potential to transform the BFSI sector by enhancing efficiency, personalization, and security. However, ethical challenges must be addressed through responsible development and deployment practices. By considering diverse perspectives and engaging with counterarguments, generative AI can contribute positively to the financial industry's future. Ensuring transparency, fairness, and collaboration between human and AI systems will be key to harnessing the full potential of this transformative technology.

Adhering to Financial Regulations
Ensuring that AI-driven solutions comply with existing financial regulations and standards is a significant challenge for financial institutions. Regulatory bodies require that any technology used in financial operations must adhere to strict guidelines to prevent malpractice and ensure consumer protection. For instance, the implementation of generative AI in JPMorgan Chase's operations involves rigorous compliance with the Dodd-Frank Act, which mandates transparent and fair practices in financial services . This means that AI systems used in trading, risk management, and customer service must be thoroughly vetted and continuously monitored to meet regulatory standards.

Data Protection Laws

Meeting the requirements of data protection laws like the General Data Protection Regulation (GDPR) is another critical aspect. GDPR mandates that personal data must be handled with utmost care, ensuring privacy and security. Financial institutions using generative AI must ensure that their systems are compliant with GDPR by implementing robust data protection measures. For example, HSBC has integrated GDPR compliance into its AI systems to ensure that customer data is anonymized and securely stored, thereby protecting it from unauthorized access and breaches.

Checklist for Deploying Ethical AI in BFSI

- **Stakeholder Engagement**: Involve diverse stakeholders, including employees, customers, and regulators, in the AI development process.
- **Human-in-the-Loop Systems**: Ensure critical AI decisions, such as loan approvals, involve human oversight.
- **Ethics Committees**: Establish ethics committees to oversee AI projects and ensure alignment with ethical standards.
- **Robust Testing and Monitoring**: Conduct extensive testing and continuous monitoring to ensure AI resilience and address technical issues promptly.
- **Data Anonymization and GDPR Compliance**: Anonymize customer data and implement robust measures to comply with GDPR and other data protection laws.
- **Transparency and Explainability**: Develop AI systems that provide clear explanations for their decisions and maintain open communication with customers.
- **Bias Detection and Inclusive Training**: Implement tools to detect and mitigate biases in AI systems and use diverse datasets for training.
- **Sustainable Practices**: Develop AI systems using sustainable and energy-efficient practices and assess their societal impact regularly.
- **Clear Accountability Framework**: Define clear roles for AI governance and establish processes for reporting AI-related incidents.
- **Reskilling and Upskilling**: Invest in training programs to help employees develop skills that complement AI technologies.

Ongoing Monitoring and Reporting

Establishing systems for the ongoing monitoring and reporting of AI activities to regulatory bodies is essential to maintaining compliance. This involves creating frameworks that allow continuous oversight and transparency of AI operations. Institutions like Capital One have developed comprehensive monitoring systems to track the performance and decision-making processes of their AI applications. These systems generate regular reports that can be audited by regulatory authorities, ensuring that the AI technologies remain compliant with financial regulations and ethical standards.

Responsible Development and Deployment of AI

Ethical AI Design

Ethical AI design involves principles and practices aimed at developing AI in a responsible manner. This includes stakeholder involvement and the establishment of ethical review boards to oversee AI projects. For example, BBVA has set up an AI ethics committee that includes representatives from various departments, such as legal, compliance, and IT, to ensure that AI development aligns with ethical standards and societal values.

AI Governance

AI governance refers to the frameworks and strategies used to govern AI development and deployment within financial institutions. These frameworks help manage the risks associated with AI and ensure that its implementation is aligned with the institution's strategic objectives and regulatory requirements. JPMorgan Chase, for example, has implemented a robust AI governance framework that includes policies for data management, algorithm transparency, and accountability. This framework ensures that AI applications are developed and deployed responsibly, minimizing risks and maximizing benefits.

Human Oversight

Human oversight in AI-driven decisions is crucial, particularly in critical financial decisions such as loan approvals and investment strategies. While AI can process and analyze vast amounts of data more efficiently than humans, final decisions often require human judgment to account for nuances and ethical considerations that AI might miss. At HSBC, for instance, AI-driven credit scoring models are used to assess loan applications, but human loan officers review the AI's recommendations before making the final decision. This combination of AI efficiency and human judgment helps ensure fairness and accuracy in financial decision-making.

Real-Time Examples from Banks and Financial Institutions

- **JPMorgan Chase**: Utilizes the COiN system to automate the review of legal documents, ensuring compliance with financial regulations such as the Dodd-Frank Act. The system is monitored continuously to adhere to regulatory standards.

- **HSBC**: Implements AI for fraud detection and customer service, ensuring GDPR compliance by anonymizing and securing customer data. AI systems are integrated with robust data protection measures to safeguard privacy.

- **Capital One**: Develops AI monitoring systems that track AI performance and decision-making processes, generating regular compliance reports for regulatory bodies.

- **BBVA**: Establishes an AI ethics committee to oversee AI projects, ensuring ethical development and deployment of AI technologies. The committee includes representatives from legal, compliance, and IT departments.

- **Commonwealth Bank of Australia (CBA)**: Uses AI-driven chatbots and virtual assistants to handle customer inquiries, improving service efficiency and allowing human agents to focus on more complex issues. AI applications are governed by policies that ensure transparency and accountability.

Automation and the Future of Work

Generative AI is transforming the BFSI sector by automating various tasks, leading to significant changes in the nature of work. AI systems can handle routine tasks such as data entry, transaction processing, and customer service inquiries, which traditionally required human intervention. This shift towards automation enhances efficiency and reduces operational costs but also raises concerns about job displacement. For example, AI-driven chatbots in the Commonwealth Bank of Australia handle a vast majority of customer inquiries, freeing up human agents to focus on more complex issues. However, the automation of these roles means that some positions may become redundant, prompting a need for workforce adaptation.

Reskilling and Upskilling

To adapt to an AI-driven workplace, BFSI professionals must embrace reskilling and upskilling initiatives. Financial institutions are increasingly investing in training programs to help employees develop new skills that complement AI technologies. For instance, JPMorgan Chase has launched various training programs focused on data science and machine learning to prepare their workforce for the future of AI integration. These programs aim to equip employees with the necessary skills to work alongside AI, such as interpreting AI outputs, managing AI systems, and ensuring compliance with regulatory standards.

Innovation vs. Ethical Constraints

Balancing Innovation with Ethical Considerations

Financial institutions face the challenge of balancing rapid innovation with ethical considerations. While generative AI offers numerous benefits, such as improved efficiency and personalized services, it also poses ethical risks, including bias, privacy concerns, and transparency issues. To innovate responsibly, banks and financial institutions must implement robust ethical frameworks and governance structures. For example, HSBC has established an AI ethics committee to oversee the development and deployment of AI technologies, ensuring that ethical considerations are integrated into every stage of the process.

Case Study

BBVA provides a compelling example of balancing innovation with ethical considerations. The bank has implemented AI-driven loan approval processes that improve efficiency while maintaining human oversight to ensure fairness and transparency. Additionally, BBVA's AI ethics committee continuously monitors AI applications to mitigate biases and ensure compliance with ethical standards. Another example is Capital One, which uses AI to personalize customer interactions while adhering to strict data privacy regulations to protect customer information.

Future Trends and Predictions

Emerging Technologies

Emerging technologies, such as quantum computing, are poised to further revolutionize the BFSI sector. Quantum computing could significantly enhance the capabilities of generative AI by enabling faster and more complex data processing. This technological advancement would allow financial institutions to develop more sophisticated models for risk management, fraud detection, and investment strategies. As quantum computing becomes more accessible, its integration with AI will likely lead to groundbreaking innovations in financial services.

Predictions for the Future

The future of generative AI in BFSI is expected to involve greater integration and sophistication of AI systems. AI technologies will become more adept at handling complex tasks, leading to even greater automation and efficiency. However, this evolution will also bring new ethical and regulatory challenges. Financial institutions must remain vigilant in addressing these issues to ensure that AI advancements benefit all stakeholders equitably. The development of comprehensive ethical guidelines and regulatory frameworks will be crucial in navigating these challenges and fostering responsible AI innovation.

Consumer Trust and Perception

Building Consumer Trust

Building consumer trust in AI-driven services is essential for financial institutions. Transparency is a key factor in this endeavor. Banks must ensure that their AI systems are explainable and that customers understand how their data is used and how decisions are made. For instance, Capital One has implemented transparency measures to explain AI-driven credit decisions to customers, thereby enhancing trust and confidence in their AI applications.

Public Perception

Public perception of AI in financial services can be influenced by various factors, including media coverage and personal experiences. Addressing public concerns about AI, such as data privacy and job displacement, is crucial for maintaining a positive perception. Financial institutions can engage in open dialogue with customers, providing clear information about the benefits and risks of AI. Educational campaigns and customer support initiatives can help demystify AI technologies and build a more informed and trusting customer base.

10.2. Framing an AI Governance Policy for Generative AI in Banking

As banks integrate generative AI into their operations, establishing a comprehensive governance framework is essential to ensure ethical, transparent, and secure deployment. The following outlines a detailed AI governance policy, leveraging industry best practices and regulatory guidelines.

Corporate Culture: Fostering Ethical AI Practices

Promoting a culture that prioritizes ethical AI development and usage is critical. Banks should implement ongoing AI literacy and training programs for all employees and establish ethics committees to review AI projects, ensuring alignment with corporate values and ethical standards. HSBC has successfully adopted this approach by forming an AI ethics committee to oversee AI applications, ensuring they meet ethical guidelines and societal expectations.

People and Skills: Investing in Talent and Continuous Learning

To adapt to AI advancements, banks must invest in acquiring AI experts and data scientists and implement continuous learning and development programs. Cross-functional teams comprising IT, compliance, risk management, and business units are essential for fostering collaboration and comprehensive oversight. JPMorgan Chase exemplifies this strategy by launching training programs focused on data science and machine learning to prepare their workforce for AI integration.

AI Governance Process Design: Robust Policies and Frameworks

Developing comprehensive policies and frameworks that govern AI usage is imperative. This includes adopting a "compliance by design" approach, integrating privacy and regulatory requirements into AI model development. HSBC's AI systems, designed to comply with GDPR, illustrate this approach by anonymizing and securely storing customer data.

Information, Data, and Sources: Ensuring Data Quality and Integrity

Stringent data governance practices are necessary to maintain data quality and integrity. Banks should utilize tools and methodologies to detect and mitigate biases in data and AI models, ensuring fair and equitable outcomes. Capital One's comprehensive data management frameworks serve as a benchmark for maintaining data integrity and ensuring reliable AI outputs.

Software and Algorithms: Regular Audits and Security Measures

Regular audits of AI algorithms ensure they function as intended and are free from biases. Robust security measures must be implemented to protect AI models from unique threats like data poisoning and illicit replication. JPMorgan Chase employs advanced cybersecurity measures to safeguard their AI models, setting a high standard for security in AI governance.

Services, Infrastructure, and Platforms: Scalable and Secure Infrastructure

Developing scalable and flexible infrastructure to support AI initiatives is crucial (Table 10-1). Ensuring seamless integration of AI systems with existing IT infrastructure

CHAPTER 10   ETHICAL DILEMMAS AND FUTURE POTENTIAL OF GENERATIVE AI IN THE FINANCIAL SPHERE

and establishing continuous monitoring and maintenance protocols are necessary steps. The Commonwealth Bank of Australia exemplifies this by using AI-driven chatbots and virtual assistants to enhance service efficiency.

Table 10-1. Infrastructure

Governance Areas	Challenges	Real-Time Examples
Ethics	AI systems should be developed, deployed, and used in ways that respect human autonomy and ensure fairness and explicability.	**HSBC:** Established an AI ethics committee to oversee AI applications, ensuring alignment with ethical guidelines and societal values.
Fairness	AI models should be free from bias to prevent unjust disadvantages to any individual or group.	**BBVA:** Regular reviews by ethics committees ensure that loan approval processes remain unbiased and equitable.
Explainability and Traceability	Banks must be able to fully explain any AI-driven decision that affects customers or other individuals who provide data.	**JPMorgan Chase.** Uses COiN (Contract Intelligence) to track and explain AI decisions from data gathering to final outcomes, ensuring transparency and accountability.
Human in the Loop	Human monitoring and review are needed to ensure AI models are performing correctly and fairly.	**HSBC:** Human officers review AI-generated loan recommendations to ensure fairness and accuracy in financial decision-making.
Data Quality	Avoiding bias and ensuring fairness of decisions depends on the accuracy and integrity of data used in AI models.	**Capital One:** Implements comprehensive data management frameworks to ensure data integrity for reliable AI outputs.
Model Security	AI models are susceptible to unique threats, such as illicit replication and data poisoning.	**JPMorgan Chase:** Employs robust cybersecurity measures to protect AI models from theft and data poisoning.
Skills	Deficits of AI-specific talent and expertise pose a challenge in building, maintaining, and overseeing AI models.	**JPMorgan Chase:** Launched training programs focused on data science and machine learning to prepare their workforce for AI integration.

(continued)

Table 10-1. (*continued*)

Governance Areas	Challenges	Real-Time Examples
Compliance	Management of customer data must fully conform to privacy regulations like GDPR.	**HSBC**: Ensures GDPR compliance by anonymizing and securing customer data within their AI systems.
Governance Structure and Management	AI models must be subject to stringent oversight and integrated into existing risk management frameworks.	**BBVA**: Integrates AI oversight with their broader risk management strategies, ensuring stringent governance practices.
Accountability	Responsibility for AI decisions must be held by humans, ultimately the boards of directors.	**HSBC**: Board members are familiarized with AI functioning to ensure high-level oversight and accountability for AI decisions.

External and Internal Drivers: Adapting to Evolving Requirements

Banks' AI governance frameworks must remain adaptable to the latest advancements in AI technology, market competition, and regulatory frameworks. Aligning AI initiatives with customer needs and business objectives should be a priority. Additionally, integrating AI use into existing risk management frameworks ensures stringent oversight and clear roles and responsibilities.

Accountability: Human Oversight and Board Responsibility

Responsibility for AI decisions should ultimately rest with human experts and the board of directors, ensuring accountability and high-level oversight. HSBC's approach, where board members are familiarized with AI functioning to oversee AI decisions responsibly, serves as a model for other institutions.

Integrating generative AI into banking operations offers transformative potential, but must be governed by robust, ethical, and secure practices. By learning from industry leaders such as HSBC, JPMorgan Chase, and Capital One and adhering to regulatory guidelines, banks can develop comprehensive AI governance policies that ensure generative AI initiatives are ethical, transparent, and secure. This policy will not only enhance operational efficiency and customer satisfaction but also uphold the trust and integrity that stakeholders expect from financial institutions.

AI Governance Framework for Implementation of Generative AI in Banking

Corporate Culture: Fostering Ethical AI Practices

- **Promote AI Ethics**: Implement a culture that prioritizes ethical AI development and usage.
- **Training Programs**: Conduct ongoing AI literacy and training programs.
- **Ethics Committees**: Establish committees to review AI projects and ensure they align with corporate values and ethical standards.

People and Skills: Investing in Talent and Continuous Learning

- **Recruit AI Experts**: Invest in acquiring AI specialists and data scientists.
- **Continuous Development**: Implement learning and development programs.
- **Cross-Functional Collaboration**: Create cross-functional teams to ensure comprehensive oversight and collaboration.

AI Governance Process Design: Robust Policies and Frameworks

- **Develop Policies**: Establish comprehensive AI governance policies and frameworks.
- **Compliance by Design**: Integrate compliance requirements into AI development.
- **Ethical by Design**: Incorporate ethical considerations from the outset of AI model development.

CHAPTER 10   ETHICAL DILEMMAS AND FUTURE POTENTIAL OF GENERATIVE AI IN THE FINANCIAL SPHERE

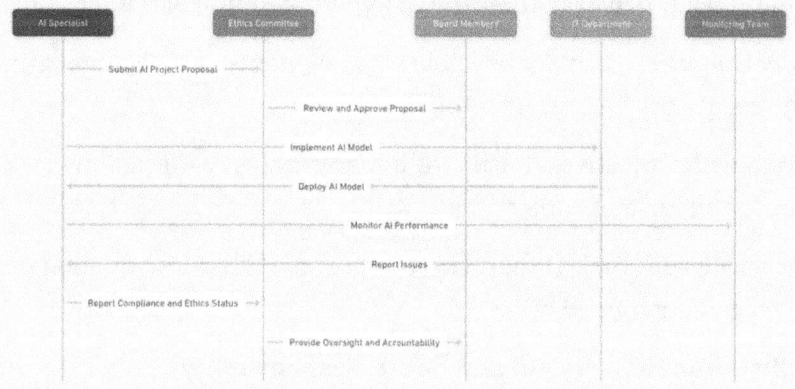

Figure 10-1. *Governance Framework*

Information, Data, and Sources: Ensuring Data Quality and Integrity

- **Data Governance**: Implement stringent data governance practices.

- **Bias Mitigation**: Utilize tools to detect and mitigate biases in data and AI models.

- **Transparency**: Ensure that AI-driven decisions are transparent and explainable.

Software and Algorithms: Regular Audits and Security Measures

- **Algorithm Audits**: Conduct regular audits to ensure AI algorithms function as intended.

- **Security Measures**: Implement robust security measures to protect AI models from threats.

- **Model Validation**: Continuously validate AI models for accuracy and reliability.

Services, Infrastructure, and Platforms: Scalable and Secure Infrastructure

- **Scalable Infrastructure**: Develop scalable infrastructure to support AI initiatives.

- **Integration**: Ensure seamless integration of AI systems with existing IT infrastructure.

- **Monitoring and Maintenance**: Establish protocols for continuous monitoring and maintenance.

External and Internal Drivers: Adapting to Evolving Requirements

- **Stay Updated**: Adapt to the latest advancements in AI technology and regulatory frameworks.
- **Align with Objectives**: Ensure AI initiatives align with customer needs and business objectives.
- **Integrate with Risk Management**: Integrate AI use into existing risk management frameworks.

Accountability: Human Oversight and Board Responsibility

- **Human Responsibility**: Ensure responsibility for AI decisions rests with human experts.
- **Board Accountability**: Board members should oversee AI decisions and be familiarized with AI functioning.
- **Consistent Standards**: Apply consistent AI standards to both internally developed and externally sourced AI applications.

This governance framework (Figure 10-1) ensures the ethical, transparent, and secure implementation of generative AI in the banking sector, leveraging industry best practices and regulatory guidelines. By fostering a culture of ethical AI practices, investing in skills and talent, and maintaining robust policies and frameworks, banks can harness the transformative potential of AI while upholding trust and integrity.

10.3. Cautionary Tales: Examples of What Could Go Wrong

Case Studies of Algorithmic Bias in Loan Approvals

Introduction

The use of AI in banking, particularly for loan approvals, has shown promise in increasing efficiency and objectivity. However, it has also raised significant concerns about fairness and bias. This case study explores specific instances where AI algorithms exhibited biases, the root causes of these biases, and their impacts on both customers and financial institutions.

CHAPTER 10   ETHICAL DILEMMAS AND FUTURE POTENTIAL OF GENERATIVE AI IN THE FINANCIAL SPHERE

Specific Instances

One prominent example of algorithmic bias in loan approvals can be traced back to the use of AI systems by several banks, including the case of Apple's Apple Card, which was accused of giving women lower credit limits compared to men (Ross, S. (2024, April 30). Major regulations following the 2008 financial crisis. In C. Clarke (Ed.), *Laws & Regulations Investing*. Fact-checked by Y. Perez. Investopedia.). Despite the New York State Department of Financial Services finding no evidence of discrimination based on sex, this case highlighted the potential for AI systems to inadvertently replicate societal biases.

Another notable instance is the historical practice of redlining in Chicago, where AI systems used for determining loan risk continued to reflect biases from past data. Even though these systems did not explicitly include race as a factor, they picked up on proxies for race from other data points, resulting in discriminatory outcomes against predominantly Black neighborhoods.

Root Causes of Bias

The root causes of biases in AI systems often stem from the data used to train these models. If the historical data reflects societal biases, the AI models trained on this data are likely to replicate those biases. For instance, if loan approval data historically favored certain demographic groups over others, the AI system would learn to favor those groups as well.

Moreover, biases can also originate from the development teams. A lack of diversity within these teams can result in unintentional biases being encoded into the AI systems. Deloitte highlights that incomplete or unrepresentative datasets can limit AI's objectivity, perpetuating a cycle of bias.

Impact and Consequences

The consequences of biased AI systems in loan approvals are profound. For the affected individuals, this bias can lead to unfair denials of loans, limiting their financial opportunities and exacerbating existing inequalities. This is particularly damaging for marginalized communities that have historically faced discrimination in lending practices.

For financial institutions, the use of biased AI systems can damage their reputation and erode trust with their customer base. In addition to reputational harm, there can also be legal and regulatory repercussions if these biases lead to non-compliance with anti-discrimination laws and regulations.

Addressing the Bias

Addressing algorithmic bias in loan approvals requires a multifaceted approach. Financial institutions need to implement robust data governance practices to ensure that the data used to train AI models is representative and unbiased. Regular audits of AI algorithms are essential to identify and mitigate biases. For example, HSBC's approach of establishing an AI ethics committee to oversee AI applications ensures that ethical considerations are integrated into the AI development process.

Involving diverse teams in the development of AI systems can also help mitigate biases. Diverse perspectives can identify potential biases that might be overlooked by a homogenous team. Continuous monitoring and validation of AI models are necessary to ensure they perform fairly and accurately over time.

Furthermore, transparency and explainability in AI systems are crucial. Banks must be able to explain how their AI systems make decisions and provide customers with clear reasons for any adverse decisions. This transparency can help build trust and allow for corrective measures to be taken when biases are detected.

The use of AI in loan approvals presents significant opportunities for enhancing efficiency and objectivity. However, it also poses substantial risks of perpetuating existing biases. By understanding the root causes of these biases and implementing comprehensive governance frameworks, financial institutions can harness the benefits of AI while ensuring fairness and equity in their lending practices. This balanced approach will help build trust with customers and comply with regulatory requirements, ultimately contributing to a more equitable financial system.

Case Study: Mastercard Accelerates Card Fraud Detection with Generative AI Technology

Mastercard, a leader in cybersecurity, has significantly enhanced its ability to detect potentially compromised cards by leveraging generative AI. This advancement comes as fraudsters continue to exploit technology to steal millions of payment card numbers through various clandestine practices. By using generative AI, Mastercard has doubled the speed at which it can detect compromised cards, thereby protecting cardholders and securing the digital ecosystem.

Fraudsters often sell stolen card numbers on illegal websites, revealing partial card details. Mastercard's new AI technology predicts the full details of these compromised cards, enabling banks to block them faster than ever before. "Until now, fraudsters may have thought they were operating in obscurity, seeking to launder the card details of

CHAPTER 10   ETHICAL DILEMMAS AND FUTURE POTENTIAL OF GENERATIVE AI IN THE FINANCIAL SPHERE

millions of unsuspecting victims. Thanks to our world-leading cyber technology, we can now piece together the jigsaw—enhancing trust to banks, their customers, and the digital ecosystem as a whole," said Johan Gerber at Mastercard.

Enhancements and Impact

Mastercard's generative AI technology scans transaction data across billions of cards and millions of merchants at unprecedented speeds. This capability allows it to identify new, complex fraud patterns and protect future transactions against emerging threats. Key benefits of this technology include

- **Doubling the Detection Rate of Compromised Cards**: By identifying potentially compromised cards more quickly, Mastercard can prevent fraud more effectively.

- **Reducing False Positives**: The technology minimizes instances where legitimate transactions are flagged as fraudulent, improving the accuracy of fraud detection by up to 200%.

- **Increasing the Speed of Identifying At-Risk Merchants**: Mastercard can now identify merchants compromised by fraudsters 300% faster, enhancing overall security in the payment ecosystem.

These enhancements enable Mastercard to alert banks more quickly and accurately when a card is likely to have been compromised. As a result, banks can block and reissue cards faster, continuously monitor attempted transactions on compromised cards, and enhance cybersecurity measures. This proactive approach instills greater trust in the digital ecosystem.

Mastercard's advancements in fraud detection build on its existing Cyber Secure suite, available since 2020. Cyber Secure uses integrated technology to create a baseline of transparent cybersecurity information on bank and merchant online profiles, including details of suspected compromised cards.

AI Failures in Fraud Detection Systems

Examples of Ineffective Fraud Detection

Despite the benefits, AI systems in fraud detection are not infallible. There have been instances where AI failed to detect fraudulent activities or erroneously flagged legitimate transactions as fraud. For example, some financial institutions have reported cases where their AI systems missed sophisticated fraud schemes due to limitations in their training data or algorithmic design.

Analysis of System Shortcomings

The shortcomings of AI systems in fraud detection often stem from the quality and representativeness of the data used to train them. Incomplete or biased datasets can lead to inaccurate predictions. Additionally, the complexity of fraud patterns can evolve faster than the AI models can adapt, leading to gaps in detection capabilities. Deloitte highlights that AI systems are only as good as the data they're trained on; unrepresentative datasets can limit AI's effectiveness and perpetuate biases.

Operational and Reputational Impact

Failures in AI fraud detection can have significant operational and reputational impacts on BFSI companies. Operationally, these failures can result in financial losses due to undetected fraud and the costs associated with false positives. Reputationally, such failures can erode customer trust and confidence. For example, a well-publicized failure in fraud detection can lead to negative media coverage and customer backlash, as seen in cases involving major banks that have experienced significant fraud incidents despite having AI systems in place.

While generative AI technology offers substantial advancements in fraud detection, it is essential to acknowledge and address its limitations. Ensuring the use of high-quality, representative data and continuously updating AI models are critical steps to enhance the effectiveness of fraud detection systems. By doing so, financial institutions can better protect their customers, maintain trust, and uphold the integrity of the financial ecosystem.

AI Bias in the Underwriting Process in the Insurance Industry

The increasing adoption of artificial intelligence (AI) in the insurance industry promises significant benefits for both insurers and insureds. AI's capability to analyze vast datasets swiftly and accurately can revolutionize risk identification, improve underwriting and claims handling, and optimize premium pricing. However, this technological advancement also introduces potential risks related to accuracy, fairness, and security. Insurers and risk professionals must understand these pitfalls to ensure that AI-driven processes do not introduce more risks than they mitigate.

Cases of Unfair Insurance Rates

AI has the potential to bring precision to actuarial models and underwriting, enabling insurers to offer tailored coverage and enhance risk management. For instance, AI can analyze diverse data sources, such as historical claims, customer behavior, litigation trends, market changes, extreme weather events, and social media posts, to create a comprehensive understanding of risk factors. This allows for better and more specific underwriting decisions and personalized insurance policies based on individual behavior and risk profiles. However, experts warn that AI can perpetuate existing biases, leading to unfair insurance pricing or denial of coverage.

Wilson Chan, CEO at Permutable AI, emphasizes the critical need to address biased data in AI systems within the insurance industry (Davies North America—The Impact of AI on Insurance Underwriting). He notes that companies often face inflated premiums and coverage restrictions due to insurers training their underwriting AI on limited or biased data. For example, AI models trained on historical flood risk data might unfairly impact companies in flood-prone areas, overlooking current climate patterns and mitigation measures, leading to higher premiums or coverage limitations.

Examination of Contributing Factors

The primary factor contributing to unfair outcomes in AI-driven insurance underwriting is the quality of the data used to train these systems. If the input data is biased, the AI's decisions will reflect those biases. Historical biases rooted in the data can adversely impact companies, especially in niche or emerging sectors where historical data may not accurately reflect current realities. Peter Wood, CTO at Spectrum Search, points out that AI systems learning from past data might assign undue risk to certain companies based on outdated criteria, leading to skewed risk assessments and higher premiums.

Another contributing factor is the lack of transparency and regular auditing of AI systems. Insurers must ensure that AI systems are trained on representative, unbiased data and that these systems are regularly reviewed and updated to eliminate biases. Transparency about the functionality of AI systems and the processes they are used in is also crucial. Chan suggests that both companies and insurers can foster a fair and responsible use of AI by adhering to measures such as unbiased data training, regular audits, and transparency in AI-driven decision-making.

Regulatory and Legal Implications

The regulatory and legal implications of biased AI systems in insurance are significant. Insurers must comply with various data protection and anti-discrimination laws, such as the European Union's General Data Protection Regulation (GDPR), the California Consumer Privacy Act (CCPA), and the U.S. Health Insurance Portability and Accountability Act (HIPAA). Ensuring compliance requires transparency, ethical AI practices, and adherence to regulatory guidelines. Failure to do so can result in legal challenges, reputational damage, and financial penalties.

For example, the historical practice of redlining in Chicago, where AI systems used for determining loan risk reflected past biases, illustrates the potential for AI to perpetuate discrimination. These biases can lead to higher insurance premiums and restrictive coverages for marginalized communities, thereby reinforcing existing disparities. Companies must engage insurers in transparent dialogues, inquire about data sets used for training AI models, and ensure mechanisms are in place to identify and mitigate biases.

To address these concerns, companies should ensure that their insurers maintain comprehensive audit trails and comply with industry standards and regulations governing AI in insurance. Regular audits, compliance checks, and adherence to ethical AI practices can help maintain trust and ensure that AI-driven decisions are fair and unbiased.

While AI has the potential to transform the insurance industry, it also introduces risks related to bias and fairness. Insurers must adopt robust data governance practices, ensure transparency in AI decision-making, and adhere to regulatory guidelines to mitigate these risks. By fostering a culture of ethical AI practices and engaging in proactive dialogues with insurers, companies can contribute to a more trustworthy and equitable insurance landscape.

AI Missteps in Investment and Trading Decisions

The adoption of artificial intelligence (AI) in financial markets has transformed investment and trading strategies, offering unprecedented speed and analytical capabilities. However, the deployment of AI has also led to significant missteps, resulting in poor financial performance and market disruptions. This case study explores instances of AI errors in financial markets, analyzes the underlying causes of these flaws, and assesses their financial and reputational impacts.

AI Errors in Financial Markets

One of the most notable cases of AI-driven investment strategy failures occurred with the "flash crash" of May 6, 2010. On this day, the U.S. stock market experienced a rapid and severe drop in prices, wiping out nearly $1 trillion in market value within minutes. High-frequency trading algorithms, which execute trades at lightning speeds, were primarily blamed for this market turmoil. The algorithms exacerbated the crash by amplifying sell orders, leading to a downward spiral of stock prices.

Another example is the 2013 incident involving Knight Capital, a financial services firm. Knight Capital's trading algorithm malfunctioned, causing the firm to incur a loss of $440 million in just 45 minutes. The error stemmed from a software glitch in the algorithm, which executed numerous erroneous trades, leading to significant financial damage and a loss of client trust.

CHAPTER 10   ETHICAL DILEMMAS AND FUTURE POTENTIAL OF GENERATIVE AI IN THE FINANCIAL SPHERE

Analysis of AI Decision-Making Flaws

The primary reasons behind AI missteps in financial markets often include overfitting, lack of market understanding, and technical errors. Overfitting occurs when an AI model is too closely tailored to historical data, failing to generalize to new market conditions. This can lead to poor performance when market dynamics shift.

For instance, during the flash crash, high-frequency trading algorithms failed to account for the sudden influx of sell orders, leading to a feedback loop that exacerbated the market decline. Similarly, Knight Capital's software glitch highlights the technical vulnerabilities inherent in complex trading algorithms. The lack of thorough testing and fail-safe mechanisms contributed to the rapid accumulation of erroneous trades.

Additionally, AI models can suffer from a lack of market understanding. While these models can process vast amounts of data, they may not fully grasp the nuances of market sentiment or unexpected geopolitical events. This gap can lead to suboptimal decision-making and increased risk exposure.

Financial and Client Trust Impact

The financial losses resulting from AI errors can be substantial. The flash crash led to temporary but massive market devaluations, impacting numerous investors and eroding confidence in automated trading systems. Knight Capital's incident not only caused a significant financial loss but also severely damaged its reputation. The firm had to seek emergency funding to cover the losses, and client trust was significantly eroded, leading to a decline in business.

Beyond the immediate financial impact, these incidents highlight the importance of robust risk management practices and the need for continuous monitoring and updating of AI models. Firms must ensure that their AI systems are equipped to handle unexpected market conditions and that adequate fail-safes are in place to prevent cascading failures.

While AI offers powerful tools for investment and trading, it also introduces new risks that must be carefully managed. Overfitting, technical errors, and a lack of market understanding are common pitfalls that can lead to significant financial and reputational damage. By learning from past mistakes and implementing rigorous testing, continuous monitoring, and robust risk management practices, financial institutions can better harness the potential of AI while mitigating its risks. Ensuring transparency and maintaining client trust are essential components of a successful AI-driven investment strategy.

CHAPTER 10 ETHICAL DILEMMAS AND FUTURE POTENTIAL OF GENERATIVE AI IN THE FINANCIAL SPHERE

Generative AI in BFSI: Navigating Data Security Challenges

In the rapidly evolving landscape of the Banking, Financial Services, and Insurance (BFSI) sector, the adoption of Generative Artificial Intelligence (GenAI) is both a boon and a bane. While GenAI offers groundbreaking opportunities to enhance customer experience, streamline operations, and introduce innovative products, it also presents significant data security challenges unique to the BFSI industry. This article explores these challenges, the distinct nature of attack vectors, compares these issues with conventional IT security practices, and discusses effective mitigation strategies and the pivotal role of the Chief Information Security Officer (CISO) in navigating this complex environment.

Understanding the Unique Attack Vectors in GenAI

GenAI systems in the BFSI sector are primarily vulnerable to three types of attacks:

1. **Data Poisoning**: Attackers inject malicious data into the training dataset, which can skew the model's output, leading to faulty decision-making.

2. **Model Inversion Attacks**: These attacks involve reverse-engineering the AI model to access sensitive data, potentially exposing the personal and financial details of customers.

3. **Adversarial Attacks**: By making subtle alterations to input data, attackers can deceive AI models into making incorrect predictions or classifications, leading to erroneous financial advice or decisions.

Each of these vectors exploits the inherent characteristics of AI models, such as their dependence on large volumes of data and their often-opaque decision-making processes.

Real-World Examples

- **Direct Attack Vector**: Imagine a malicious actor using a generative AI model to extract sensitive information. For instance, if the model is tricked into revealing credit card numbers by manipulating prompts, it demonstrates a severe breach.

- **Indirect Attack Vector**: In a finance setting, if an individual manipulates the AI prompt to exclude certain account transactions, it could facilitate embezzlement. This type of prompt injection can falsify outputs, leading to financial losses.

How GenAI Security Differs from Conventional IT Security

The security challenges posed by GenAI are fundamentally different from those encountered in traditional IT security in several ways:

- **Opacity of AI Decision-Making**: Unlike conventional IT systems, the decision-making process in AI can be less transparent, making it difficult to diagnose and rectify security breaches.

- **Scale and Sensitivity of Data**: BFSI enterprises handle highly sensitive data at a massive scale, increasing the risk and potential impact of data breaches through GenAI systems.

- **Dynamic Nature of Threats**: The adaptive nature of AI systems means they continuously learn and evolve, which can inadvertently learn from malicious inputs unless properly safeguarded.

Mitigation Strategies for BFSI Enterprises

To safeguard against these vulnerabilities, BFSI enterprises need to employ robust mitigation strategies that go beyond traditional IT security measures:

1. **Robust Data Governance and Management**: Ensuring data integrity by implementing stringent controls over data collection, processing, and storage to prevent data poisoning.

2. **Adversarial Training**: Incorporating adversarial examples into the training process to improve the model's resilience to adversarial attacks.

3. **Regular Model Auditing**: Conducting periodic audits of AI models to assess and enhance their security and ensure compliance with regulatory requirements.

4. **AI Security-Specific Regulatory Compliance**: Adhering to frameworks and guidelines specifically designed for AI security in the financial sector, such as those suggested by financial regulatory authorities in India (e.g., Personal Data Protection Bill, Niti Ayog's paper on Responsible AI for India).

Data Privacy and Security Breaches Linked to AI

Instances of Data Breaches

Several high-profile cases highlight how AI systems have led to or failed to prevent significant data breaches in the BFSI sector. For instance, Capital One's 2019 data breach exposed over 100 million customers' personal information due to a misconfigured firewall that AI monitoring tools failed to detect.

Analyzing the Role of AI in These Breaches

AI systems can inadvertently contribute to security issues due to their complexity and the vast amounts of data they handle. In the Capital One breach, the AI system's inability to detect configuration issues in real time allowed the breach to go unnoticed until significant damage had been done.

Legal and Financial Ramifications

The legal and financial consequences of such breaches are substantial. Capital One faced significant fines and legal fees, alongside a severe loss of customer trust and damage to its brand. Compliance with regulations like the GDPR and CCPA becomes even more critical in the context of AI-driven data breaches.

Generative AI in the BFSI sector offers immense potential but also poses unique data security challenges. By understanding these challenges and implementing robust mitigation strategies, BFSI enterprises can better navigate the risks associated with AI. Ensuring transparency, continuous monitoring, and adherence to ethical standards are crucial for the successful deployment of AI in the financial sector. The role of the CISO is pivotal in steering these efforts, ensuring that the adoption of GenAI enhances security rather than undermines it.

10.4. The Transformative Potential of Generative AI in Banking

At the dawn of the digital revolution, Bill Gates famously asserted, "Banking is necessary; banks are not." This statement, made in 1994, suggested a future where traditional banking institutions might be rendered obsolete by digital advancements. Three decades later, while the core functions of banking remain largely unchanged, the methods have evolved significantly. The real disruption may finally be on the horizon with the advent of Generative Artificial Intelligence (GenAI), which promises to revolutionize the industry.

CHAPTER 10   ETHICAL DILEMMAS AND FUTURE POTENTIAL OF GENERATIVE AI IN THE FINANCIAL SPHERE

The Evolution of Banking with GenAI

Generative AI, unlike previous technological innovations, has the potential to transform nearly every aspect of banking. Although the fundamental functions of banking—such as safeguarding deposits and providing loans—are likely to remain unchanged, the way these services are delivered will undergo radical transformation.

Enhancing Customer Experience and Operations

Generative AI offers banks the ability to enhance customer experiences dramatically. For instance, AI can personalize customer interactions, providing tailored financial advice based on real-time data analysis. This level of personalization can significantly improve customer satisfaction and loyalty.

Moreover, GenAI can streamline operations by automating routine tasks. For example, AI-driven chatbots can handle a vast array of customer service inquiries, freeing up human employees to focus on more complex issues. This not only improves efficiency but also allows for a more responsive and agile customer service experience.

Revolutionizing Risk Management and Compliance

One of the most promising applications of GenAI in banking is in risk management. AI algorithms can analyze vast datasets to detect patterns indicative of fraudulent activities or potential compliance issues. For example, generative AI can enhance Anti-money Laundering (AML) efforts by identifying suspicious transactions more accurately and swiftly than traditional methods.

According to a study by Accenture, the impact of generative AI on banking could be profound, with significant potential for revenue growth. The study highlights how AI can transform roles within banks, enhancing productivity and enabling employees to focus on more strategic tasks.

Unique Challenges of GenAI in Banking

Despite its potential, the integration of GenAI in banking also presents unique challenges, particularly in terms of data security and ethical considerations.

Data Security Concerns

Generative AI systems in the BFSI sector are particularly vulnerable to specific attack vectors:

1. **Data Poisoning**: Attackers may introduce malicious data into training datasets, skewing AI outputs and leading to faulty decision-making.

2. **Model Inversion Attacks**: These involve reverse-engineering AI models to access sensitive data, potentially exposing customer details.

3. **Adversarial Attacks**: By subtly altering input data, attackers can deceive AI models, resulting in incorrect predictions or classifications.

As per Gartner, both direct and indirect attack vectors pose significant threats. For example, a generative AI model could be tricked into revealing sensitive information, or manipulated prompts could cause the AI to overlook fraudulent transactions.

Ethical and Transparency Issues

The opaque nature of AI decision-making processes poses another challenge. Unlike traditional IT systems, AI decisions can be less transparent, making it difficult to diagnose and rectify issues. This opacity can lead to ethical concerns, especially if AI systems inadvertently reinforce biases present in their training data.

Effective Mitigation Strategies

To address these challenges, banks need to implement robust mitigation strategies:

1. **Robust Data Governance**: Ensuring data integrity through stringent controls over data collection, processing, and storage.

2. **Adversarial Training**: Incorporating adversarial examples into the training process to enhance AI resilience against attacks.

3. **Regular Model Auditing**: Conducting periodic audits to assess AI model security and compliance.

4. **Regulatory Compliance**: Adhering to AI-specific regulatory frameworks, such as those suggested by financial authorities, to ensure ethical use of AI.

Generative AI holds immense potential to transform the banking industry, offering opportunities for enhanced customer experiences, streamlined operations, and improved risk management. However, these benefits come with significant challenges, particularly in data security and ethical considerations. By understanding these challenges and implementing robust mitigation strategies, banks can leverage the full potential of GenAI, ensuring that the technology enhances rather than undermines their operations. The CISO's role is pivotal in this process, guiding banks through the complex landscape of AI integration while safeguarding against risks.

10.5. Embracing Digital Transformation in Banking

The pace of change in the financial services industry is relentless. Banks are accelerating their digital transformation to meet customer expectations and capitalize on new opportunities. The convergence of digital public infrastructure and breakthrough technologies is creating a unique chance for banks to strengthen customer relationships. Concepts like Banking as a Service (BaaS) are expected to play a crucial role in helping banks stay relevant and set new performance standards.

Globally, banks are becoming increasingly dissatisfied with their legacy core banking systems. They are adopting partnership models with Fintechs and tech companies to upgrade their technology architecture and remain competitive. Over the past year, many customers have opted for financial products from alternative providers instead of their main banks, highlighting the need for traditional banks to embrace technological advancements and adopt a tech-like mindset.

Traditional banks can learn from tech startups by fostering a culture of innovation and agility. Banks need to become more agile, quickly test and scale new opportunities, and continuously assess new opportunities to remain competitive. Banks should adopt a culture that accepts innovation failures to foster creativity. Viewing Fintechs as partners rather than vendors is crucial, emphasizing collaborative input. Additionally, aligning innovation teams with profit and loss (P&L) responsibilities ensures a unified approach to innovation and business objectives.

Technology can significantly enhance customer experience in banking. Many banks are already leveraging technology to provide exceptional customer experiences, particularly in retail banking. There is a substantial opportunity to digitize customer experiences within the SME and corporate segments, meeting their financial needs more effectively. Banks can foster innovation by partnering with Fintechs, adopting a frugal mindset, and simplifying processes. Fintechs often display a hunger for innovation that can be less pronounced in larger organizations. By focusing on simplicity and quick iteration, banks can emulate the rapid innovation cycles of tech companies.

Separating core and non-core technology is essential for balancing security and agility. Banks should fortify security and compliance for core systems while allowing more flexibility and experimentation with non-core technologies. This approach enables rapid innovation without compromising security. The competitive landscape is evolving towards increased collaboration between banks and Fintechs within the BaaS model. By adopting a rapid iteration approach and leveraging BaaS, banks can explore numerous opportunities at a low cost, achieving swift time-to-market while maintaining control over customer relationships.

For a service-oriented approach through BaaS, choosing the right partners and clear communication are vital. Banks should focus on delivering innovative products and services to untapped customer segments while maintaining strong customer relationships during the transition. Embedded finance, where banks integrate financial services into non-financial platforms, presents both challenges and opportunities. Partnering with BaaS providers allows banks to engage with multiple platforms quickly and identify viable opportunities, enhancing their ability to offer embedded finance solutions.

A combination approach is recommended for banks deciding between building their core technology and buying new-age tech solutions. Banks should focus on building technologies that provide a competitive edge while partnering with Fintechs for other solutions, allowing them to be agile and cost-effective. Rapid execution and adaptability are crucial for banks aiming to emulate tech disruptors. Banks should try numerous opportunities, launch use cases quickly, and refine them in real time, contrasting with traditional banks' prolonged product launch cycles.

As AI continues to revolutionize the BFSI sector, it is imperative that financial institutions adopt a responsible approach to its integration. This involves addressing ethical challenges, learning from past mistakes, and implementing robust governance frameworks. By examining real-world examples and the progress made by the tech industry, this chapter underscores the importance of balancing innovation with ethical considerations. The nuanced expectations of the BFSI sector from tech companies highlight the need for tailored solutions that ensure fairness, transparency, and security. As we move forward, the role of governance frameworks in shaping AI's future in the financial industry cannot be overstated. Through responsible AI practices, the BFSI sector can harness the full potential of generative AI, driving innovation while upholding the trust and integrity essential to financial services.

10.6. Summary

1. Generative AI in the BFSI sector is transforming operations, from fraud detection and document processing to personalized customer interactions and loan approvals, demonstrating its broad utility and impact.

2. While GenAI offers significant benefits, it also poses risks such as perpetuating biases, particularly in credit scoring and loan approvals, necessitating robust bias detection and mitigation strategies to ensure fairness and equity.

3. The handling of vast amounts of customer data by AI systems raises serious privacy concerns, making compliance with data protection regulations like GDPR crucial, alongside the need for transparency in AI decision-making.

4. Financial institutions must adhere to strict regulatory requirements, ensuring that AI-driven solutions meet financial regulations and data protection standards through continuous monitoring, ethical AI design, and human oversight.

5. The automation brought by AI can lead to job displacement, highlighting the importance of reskilling and upskilling initiatives to prepare the workforce for an AI-driven environment.

6. Algorithmic bias in AI systems has led to unfair outcomes in loan approvals and insurance underwriting, underscoring the need for diverse development teams, robust data governance, and regular audits to mitigate these biases.

7. Emerging technologies like quantum computing are expected to further enhance the capabilities of AI in the BFSI sector, enabling more sophisticated risk management and fraud detection strategies.

8. Building consumer trust in AI-driven financial services is critical, which requires transparency and clear communication about how AI systems make decisions and use customer data.

9. The chapter presents real-world examples of both successful AI implementations in financial institutions and notable failures, illustrating the need for continuous improvement and vigilance in AI deployment.

10. The integration of Generative AI in banking holds immense potential, but it must be guided by ethical considerations and robust governance frameworks to ensure that innovation does not compromise fairness, security, and trust.

References

1. Shannon, C. E. (1948). A Mathematical Theory of Communication. Bell System Technical Journal, 27, 379-423.

2. Goodfellow, I., Pouget-Abadie, J., Mirza, M., Xu, B., Warde-Farley, D., Ozair, S., ... & Bengio, Y. (2014). Generative Adversarial Nets. Advances in Neural Information Processing Systems, 27, 2672-2680.

3. Kingma, D. P., & Welling, M. (2013). Auto-Encoding Variational Bayes. arXiv preprint arXiv:1312.6114.

4. Vaswani, A., Shazeer, N., Parmar, N., Uszkoreit, J., Jones, L., Gomez, A. N., ... & Polosukhin, I. (2017). Attention is All You Need. Advances in Neural Information Processing Systems, 30, 5998-6008.

5. Brown, T. B., Mann, B., Ryder, N., Subbiah, M., Kaplan, J., Dhariwal, P., ... & Amodei, D. (2020). Language Models are Few-Shot Learners. Advances in Neural Information Processing Systems, 33, 1877-1901.

6. Kingma, D. P., & Welling, M. (2019). An Introduction to Variational Autoencoders. Foundations and Trends® in Machine Learning, 12(4), 307-392.

7. Dosovitskiy, A., Beyer, L., Kolesnikov, A., Weissenborn, D., Zhai, X., Unterthiner, T., ... & Houlsby, N. (2020). An Image is Worth 16x16 Words: Transformers for Image Recognition at Scale. arXiv preprint arXiv:2010.11929.

APPENDIX A

Glossary of Key Terms

- **Autoencoder:** A type of neural network used to learn efficient data representations in an unsupervised manner.

- **Generative Adversarial Network (GAN):** A framework for training generative models through adversarial processes involving a generator and a discriminator.

- **Transformer:** A neural network architecture based on self-attention mechanisms, widely used in NLP and other fields.

- **Reinforcement Learning from Human Feedback (RLHF):** A technique where models learn from human-provided feedback to align with human values and preferences.

- **Direct Preference Optimization (DPO):** An optimization approach that directly uses preference data to adjust model behavior without a separate reward model.

- **Differential Privacy:** A method for ensuring individual privacy in data analysis by adding noise to the data or results.

- **Federated Learning:** A decentralized approach to training models where data remains on local devices and only model updates are shared.

APPENDIX B

List of Key Figures and Contributions

- **Claude Shannon:** Pioneering work in information theory
- **Ian Goodfellow:** Introduction of Generative Adversarial Networks (GANs)
- **Diederik P. Kingma:** Development of Variational Autoencoders (VAEs)
- **Ashish Vaswani:** Introduction of the Transformer architecture
- **OpenAI:** Development of the GPT series and advancements in large-scale pre-trained models

Index

A

A/B testing, 171
Accountability, 38, 185, 232, 291, 298, 299, 305, 308
Advanced biometrics, 228
Advanced data analytics, 210
Advanced data extraction tools, 245
Adversarial attacks, 316, 320
Adversarial training, 320
AI, *see* Artificial intelligence (AI)
AI2, *see* Allen Institute for AI (AI2)
AI algorithms, 9, 100, 101, 131, 140, 146, 153, 185
AI-assisted tax planning tools, 173
AI-driven approach, 104–106, 134
AI-driven chatbots, 100, 296, 300, 304, 319
AI-driven credit scoring models, 299
AI-driven customer service, 286
AI-driven data analytics, 243
AI-driven innovations, 126–128, 178
AI-driven investment strategies, 268, 269, 314
AI-driven personalization, 11, 145, 285
AI-driven predictive analytics tools, 179
AI-driven predictive models, 150, 296
AI-driven systems, 93, 110, 120, 124, 135, 149
AI-driven tools, 12, 102, 126, 130, 136, 143, 145, 165, 173, 176, 197, 200
AI-generated content, 4, 6, 10
AI governance policy
    accountability, 305
    AI governance process design, 303
    corporate culture, 303
    external and internal drivers, 305
    governance framework, 307
    information, data and sources, 303
    infrastructure, 304
    people and skills, 303
    services, infrastructure and platforms, 303
    software and algorithms, 303
AI-Human collaboration, 130
AI integration, challenges
    advanced training for technical teams, 281
    AI skill development, 282
    continuous learning and development, 283
    employee training in AI, 281
    market and customer-related challenges, 284–288
    measure training effectiveness, 283
    partnerships with educational institutions and technology firms, 283
    role-based training approach, 280, 282
    skill and knowledge gaps, 281
    training for non-technical staff, 281
    workforce development, 280
AI-powered chatbots, 90, 95, 109, 127, 133, 144, 178, 179

INDEX

AI-powered digital learning, 90
AI-powered financial planners, 46
AI-powered tools, 9, 102
AI-powered virtual assistants, 178
AI solutions, 242, 248, 249
Algorithmic bias, 174, 177, 308–310
Algorithmic trading, 4, 9, 11, 124, 149
Allen Institute for AI (AI2), 29, 32
Allianz SE, 74–76
AlpacaEval, 29
AlphaSense, 152
Amazon Web Services (AWS), 46, 110
AML, *see* Anti-money laundering (AML)
Anomaly detection, 135, 150
Ant Financial, 180
Anti-money laundering (AML), 89, 100, 106, 137, 319
AR, *see* Augmented reality (AR)
Artificial intelligence (AI), 58, 81, 123, 130, 202
    advancements, 135
    assessment and strategy formation, 242–243
    building consumer trust, 302
    building LLMs, 45
    compliance systems, 152
    decision-making flaws, 315
    errors, 314, 315
    ethics, 37, 139, 299, 301, 306
    financial industry, 128
    fraud detection and security measures, 150, 151
    governance framework, 306–308
    governance process design, 303, 306
    investment advice, 153
    portfolio management, 147
    predictive analytics, 152
    predictive trading systems, 149
    real-time data processing, 152
    skill development, 262, 282
    technologies, 126
    trading strategies, 148, 149
    transformative force, 147
Asset allocation, 131, 141, 148, 170
Audit functions, 107–109
Augmented reality (AR), 58, 59, 65, 67, 68, 72–74, 228
Autoencoders, 26, 33, 54
Automated compliance reporting tools, 233
Automated decision-making, 20, 270
Automated investment platforms, 286, 288
AWS, *see* Amazon Web Services (AWS)
AWS CloudWatch, 172
Ayasdi, 128, 152

B

BaaS, *see* Banking as a Service (BaaS)
Backpropagation algorithm, 3, 5, 23
Backtesting, 171
Balanced approach, 15, 128, 129
Bank business support functions
    data privacy and security, 102
    finance department, 101
    legal domain, 102
    recruitment tools, 102
Banking as a Service (BaaS), 321, 322
Banking, Financial Services, and Insurance (BFSI), 10, 11, 57, 85
    challenges, 81
    data, 81
    digital transformation, 80
    FinTech, 57
    future, 58
    landscape and envision, 86

technological advancements, 57, 58
technological changes, 81
Banking industry, 81, 99, 109
    AI, 58
    blockchain, 58
    challenges, 59–64
    data analysis, 60
    digital and mobile banking, 58
    ML, 58
    RegTech, 58
    technologies, 59, 60
Banking modernization
    compliance capabilities, 89
    data analytics, 88
    flexibility and customization, 89
    flexible banking operations, 90
    operations, 88
    tools, 89
    transactions, 90
Banking operations, 58, 81, 96, 99, 305
Bank of America, 94, 99, 109, 126, 127, 133, 144, 179
BBVA, 296, 299–301, 304, 305
Behavioral analytics, 171
BERT, *see* Bidirectional Encoder Representations from Transformers (BERT)
Betterment, 136, 140, 143
BFSI, *see* Banking, Financial Services, and Insurance (BFSI)
Biases, 4, 17, 18, 28, 31, 36, 41, 146, 156, 185, 232, 245, 246, 251, 253, 263, 268, 309, 310
Bidirectional Encoder Representations from Transformers (BERT), 27, 40
Big data analytics, 209
BlackRock, 126, 131, 152
BlackRock's Aladdin platform, 102, 141, 152
Blockchain, 57, 58, 65, 67, 68, 72–75, 81, 154, 155, 202, 228
    for data security, 213
    integration, 178
    for secure data sharing, 212
    technology, 207

C

California Consumer Privacy Act (CCPA), 214, 229, 313
Capital One, 132, 151, 257, 296, 299–305, 318
Card fraud detection, Mastercard, 310, 311
CBA, *see* Commonwealth Bank of Australia (CBA)
CBS, *see* Core banking systems (CBS)
CCPA, *see* California Consumer Privacy Act (CCPA)
Challenges
    in BFSI Industry, 269
    personalized insurance models
        potential challenges and barriers, 214
        regulatory hurdles, 214
        strategies for overcoming challenges, 214, 215
Change management
    Kotter's 8-step process, 276–278
    Kurt Lewin's three-stage model, 274
        changing, 275
        refreezing, 275, 276
        unfreezing, 275
    McKinsey's 7-S framework, 278
        shared values, 279
        skills, 279

INDEX

Change management (*cont.*)
    staff, 280
    strategy, 279
    structure, 279
    style, 279
    systems, 279
Change management, AI adoption, 251, 254
    checklist for implement generative AI, 254–257
    cultural shifts, 254
    employee engagement and communication, 254
    external change management, 252, 255
    internal change management, 252, 254
    key questions for management, 252, 253
    manage customer expectations and experience, 255
    navigate integration of generative AI, 251, 252
    partnerships and ecosystem collaboration, 256
    regulatory compliance and stakeholder communication, 256
    training and upskilling, 252
Chatbots, 29, 58, 93, 95, 100, 111–113, 133, 144, 170
ChatGPT, 28, 32, 130
Chronic disease management, 221, 226
CI/CD, *see* Continuous Integration/Continuous Deployment (CI/CD)
Citadel Securities, 134
Client communication, 125, 170
ClimateAI, 183
Climate change-related insurance products, 225
    agricultural insurance, 225
    catastrophe insurance, 225
    parametric insurance, 225
Climate risk insurance
    agricultural insurance, 220
    catastrophe insurance, 220
    generative AI and LLMs, 221
    parametric insurance, 220
Cloud computing, 60, 61, 81, 96, 231, 232, 236, 248
Cloud platforms, 46, 109, 110
COiN, *see* Contract Intelligence (COiN)
Collaborative approaches, 39, 139, 150
Collaborative efforts, 139, 185, 186
Commonwealth Bank of Australia (CBA), 296, 300, 304
Community-driven projects, 53
    architecture, 33
    collaborative and decentralized approaches, 32
    fairness and mitigating bias, 31
    interpretability, 32
    journey, 33
    OpenAssistant, 30
    open models, 30
Competitive edge, 243, 245, 247, 262
Complex regulatory landscape, 229
"Compliance by design" approach, 235, 303, 306
Compliance monitoring, 105, 107, 232
Comprehensive change management, 252
Comprehensive governance framework, 243, 302
Consumer education and marketing, 227
Continuous innovation, 73, 268, 270
Continuous Integration/Continuous Deployment (CI/CD), 171
Continuous monitoring, 174, 176, 248, 257–259

INDEX

Continuous training and development, 234, 249
Contract Intelligence (COiN), 99, 102, 126, 296, 300, 304
Core banking systems (CBS), 86, 87
Corporate culture, 303, 306
CORS, *see* Cross-Origin Resource Sharing (CORS)
Cost management, 60, 61, 64, 71, 75
Creativity, 5, 23
Credit risk model drift, Global Bank Inc., 270, 271
Critical ethical issues, AI-driven financial services
   accountability, 185
   biases, 185
   privacy, 184
Cross-Origin Resource Sharing (CORS), 157
Customer-centric insurance models, 208, 209
Customer-centric solutions
   customer satisfaction and engagement, 146, 147
   privacy and security concerns, 146
Customer engagement, 103, 109, 145, 209, 210
Customer interaction channels, 208
Customer retention and growth, 60–63
Cutting-edge AI innovations, 179
   blockchain integration, 179
      IBM and Maersk's TradeLens, 180
   NLP for customer interactions, 179
      Bank of America's Erica, 179
   predictive analytics tools, 179
      Kensho technologies, 179
Cutting-edge developments, 14
Cutting-edge technologies, 201

Cyber insurance policies, 219, 224
Cybersecurity, 59, 61, 63, 65, 67, 68, 70, 73–76
Cybersecurity regulations, 230
Cybersecurity risks, 66

D

Dashboard metrics, 258
Data acquisition and quality, 267, 268
Data bias, 155, 156, 218, 289
Data breaches, 219, 318
Data cleaning, 39–40, 47
Data collection, 243
   high-quality data, 245, 246
   and management, 244
   transparency, GenAI at HSBC, 245–247
Data-driven decision-making, 103
Data engineering, 47
Data integration, 48
Data lifecycle management, 246
Data management, 60, 62, 63, 66, 69, 70, 76
Data poisoning, 316, 320
Data privacy and security, 285
Data privacy challenges, 288
Data protection laws, 298
Data Quality Assurance, 245
Data security challenges, 316
Data security concerns, 288, 320
Decentralized Finance (DeFi), 181
Decision Intelligence, 150
Deep learning, 6, 26, 135
DeFi, *see* Decentralized Finance (DeFi)
Differential privacy, 37
Digital innovation, 68, 75, 201, 203
Digital insurance, 229

INDEX

Digital platforms, 57, 59, 62, 64, 66, 69, 72, 75
Digital transformation, 59, 61, 66, 73, 201, 321, 322
    catalysts, 202
    challenges and considerations, 203
    initiatives, 231
    in insurance sector, 202
    telematics in auto insurance, 203
Direct attack vector, 316
Direct Preference Optimization (DPO), 28–33
Discrimination, 218
Dodd-Frank Act, 268, 269, 297, 300
Domain-specific skills, 261
DPO, *see* Direct Preference Optimization (DPO)

E

Effective change management, 15, 252
Effective data management strategies, 243, 246
Emerging technologies, 59, 65, 67, 68, 270, 301
eMoney Advisor, 173, 176
Emotional intelligence, 22
Enhanced customer engagement, 13
Enterprise integration checklist, 78–80
Erica, 126, 127, 133, 144
Ethical AI design, 299
Ethical and fair AI, 22
Ethical and regulatory compliance, 263, 264
Ethical considerations, 14, 297
    adhere financial regulations, 297
    balancing innovation, 301
    data protection laws, 298
    ongoing monitoring and reporting, 299
Ethics committees, 298, 306
Evolution, generative AI, 5, 6
Evolution of Banking, 319
Explainability, 263, 310
Explainable AI (XAI), 137, 142, 185, 285
External change management, 252, 255

F

Federated learning, 37, 45, 56
Feedback loops, 113, 258, 259, 270
Few-shot prompt, 195, 196
FICO, 153
Fidelity Go, 173
Financial and client trust, 315
Financial industry
    AI, 137
    challenges and ethical considerations, 153, 154
    investment banking and trading, 123
    LLMs, 132
Financial institutions, 66, 139, 146, 154
    Ant Financial, 180
    upstart leverages AI, 180
    Zest AI, 180
Financial NLP, 48
Financial planning and advisory
    decision-making in wealth management, 176
    evolve regulatory framework, 177
    focus on value-added services, 176
    long-term financial planning, 176
    predict future financial outcomes, 176
    regulatory landscape and compliance, 177
    strategies for navigate challenges, 177
    technical and ethical challenges, 177

traditional client-advisor
relationship, 176
transformative role of AI, 175
Financial planning and advisory services
AI, 137
clients-advisor interactions, 137
collaboration, 139
ethical considerations, 139
financial advice, 136
financial systems, 139
future, 138
global financial regulations, 137
innovations and potential
disruptions, 138
long-term strategies and wealth
management, 136
responsible AI development, 139
streamline compliance processes, 138
transparent AI models, 137
XAI, 137
Financial planning applications, 50
Financial planning system, 47
Financial planning tasks, 49
Financial services, 64, 166
AI, 65
blockchain, 65
challenges, 66–68
competitive performance, 69
cost management, 69, 71
customer convenience, 69
customer experience and
engagement, 71
cybersecurity, 65, 68, 70
data management, 69, 70
digital innovation, 68
digital revolution, 64
fraud detection and prevention, 68,
69, 71, 81

institutions, 65
market agility, 68
market volatility, 69, 71
ML, 65
operational efficiency, 69
operational stability, 68
personalization, 65
privacy concerns, 69
providers, 67
quality service delivery, 69
RegTech, 65
regulatory compliance, 68, 70
risk management, 69, 71
stable growth, 69
talent acquisition and retention, 69, 71
technological integration, 68, 70
technologies, 65, 67, 68
Fine-tuning, 44, 51, 112
FinTech, 57, 58, 86, 138, 321, 322
Flowcharts, 258
Forrester Research, 92
Forward-looking planning, 249
Fraud detection, 106, 119, 120, 300,
301, 310–312
and prevention, 60, 62, 64, 67, 68,
71, 75, 77
and risk management, 155, 156
and risk mitigation, 166
Future of generative AI, 302

G

Gamification, 145
GANs, *see* Generative adversarial
networks (GANs)
Gantt charts, 258
GDPR, *see* General Data Protection
Regulation (GDPR)

337

INDEX

GenAI, *see* Generative AI (GenAI)
GenAI security
    vs. conventional IT security, 317
GenAI systems
    types of attacks, 316
General Data Protection Regulation (GDPR), 138, 154, 214, 215, 229, 298
Generative adversarial networks (GANs), 3, 6, 26
    applications, 35
    architecture and training process, 34
    differential privacy, 37
    domains, 36
    ethical considerations, 37
    privacy concerns, 37
    self-attention mechanism, 35
    series, 36
    transformers, 35
Generative AI (GenAI), 25, 85, 93, 101, 201, 202, 245, 295
    adds value, 80
    application, 95
    autoencoders, 26
    BERT, 27
    BFSI, 80
    bias and fairness, 18
    challenges and future directions, 16, 17
    chatbots, 95
    ChatGPT, 28
    deep learning, 26
    domains, 27
    DPO, 28
    enterprise setting, 78
    ethical considerations, 4
    ethical dilemmas and future potential, 16
    evaluation, 29
    evolution, 3, 5, 6
    future, 53
    GANs, 26
    Google Cloud, 97
    governance frameworks, 16
    handling transactions, 94
    integration, 77
    IT infrastructure and applications, 98
    IT infrastructure management, 96
    IT restructuring, 98
    IT security, 98
    language intelligence technologies, 17, 18
    Markov chain, 26
    mathematical theory of communication, 25
    models, 36, 202, 203
    modernization, 97
    MT-Bench, 29
    neural networks, 26
    online KYC processes, 94
    OpenAI, 27
    Open LLM, 30
    operations, 99
    for personalized health insurance policies, 215–217
    privacy concerns, 18
    RLHF, 28
    roadmap, 242
    rule-based approaches, 3
    security, 100
    systems, 320
    transformative impact, 10
        challenges and potential applications, in BFSI, 10
        claims processing, 9
        credit assessment, 9

innovations in investment
banking, 11, 12
regulatory adaptations, 9
regulatory compliance, 9
transactional activities, 9
transformative practices in modern
financial services, 12–14
transforming banking, 11
transformers, 27
Zephyr model, 29
Generative Pre-trained Transformer
(GPT), 27
Geolocation technology, 222
GFIN, *see* Global Financial Innovation
Network (GFIN)
Global Bank Inc., 270, 271
Global Financial Innovation Network
(GFIN), 186
Goldman Sachs, 68–70, 134
GPT, *see* Generative Pre-trained
Transformer (GPT)

H

HFT, *see* High-frequency trading (HFT)
Hidden Markov Models, 8
High-frequency trading (HFT), 127, 134,
314, 315
Hiring, 260, 262
HSBC, 142, 144, 151, 155
Human-in-the-Loop systems, 298
Human oversight, 299

I

IBM Watson, 152, 155
Inclusivity, 19, 21, 22

Indirect attack vector, 316
Industry collaboration, 139
Innovation and product development, 227
Innovative AI models, 12
Insurance industry, 72
blockchain, 74
challenges, 72–74
cost management, 75
customer experience and
engagement, 75, 77
customer trust, 75
cybersecurity, 74–76
data management, 76
data utilization, 75
digital innovation, 75
ensuring thoroughness, 75
fraud detection and prevention, 75, 77
insurers, 72, 74
IoT, 72
market volatility, 75, 77
predictive analytics, 72
privacy concerns, 75
regulatory compliance, 74, 76
resource allocation, 76
risk management, 75, 77
stable growth, 75
streamline claims processing, 75, 76
talent acquisition and
retention, 76, 77
technologies, 76
traditional values, 75
Insurance product innovation, 228
AI and ML algorithms, 205
approaches to insurance, 204, 205
big data in risk analysis, 204
blockchain technology, 207
digital innovation, 203

INDEX

Insurance product innovation (*cont.*)
    digital transformation, 201–203 (*see also* Digital transformation)
    implications for insurers, 207
    implications for policyholders, 207, 208
    integration, IoT devices, 206, 207
    traditional risk assessment methods, 203
InsurTech startups, 72–74
Integrating and scaling generative AI
    challenges and opportunities, 249
    key questions for management, 250, 251
    organizational adaptation, 248
    organizational scalability, 249
    scalability, 248
    strategic approach, 247
    strategies for successful AI integration, 249
    technical scalability, 248
Intelligent chatbots, 100
Intelligent portfolios, 136
Internal change management, 254
Internet of Things (IoT), 59, 65, 72–74, 142, 202
    devices, 142, 206
        dynamic pricing models, 206
        encourage risk-reducing behaviors, 206
        personalized risk profiles, 206
        real-time data processing, 206
        risk segmentation, 207
    for real-time data collection, 212
Investment banking and trading, 123, 165
    adopters and success stories, 126
    AI, 123, 125, 128, 132, 165
    AI-driven systems, 124
    AI-Human collaboration, 130
    algorithmic trading, 124
    anomaly detection algorithms, 135
    asset allocation, 131
    balanced approach, 129
    client communication, 125
    compliance processes, 128
    customer satisfaction and engagement, 133
    customer service, 127
    decision making, 124
    deep learning, 135
    enhance returns, 132
    factors, 124
    financial markets, 129
    financial products, 125
    financial services, 124
    fraud detection and risk mitigation, 127, 135
    HFT, 134
    investment strategies, 131
    LLMs, 124, 125
    market trend forecasts, 134
    operational efficiency, 125
    portfolio management, 126, 130
    portfolio optimization, 125
    predictions, 133
    professionals, 129
    regulatory compliance, 125
    risk management, 131
    sentiment analysis, 125
    trading strategies, 127
    traditional practices, 147
    transformative potential, 126
    vast datasets, insights, 134
IoT, *see* Internet of Things (IoT)

J

JD Power's Banking Satisfaction Study, 146
JPMorgan Chase, 126, 135, 154, 243, 299

K

Kensho, 149
Key Performance Indicators (KPIs), 257, 262
Know Your Customer (KYC), 93, 106
Kotter's 8-Step Process, 272
Kullback-Leibler (KL), 34

L

Language intelligence, 7, 8, 17–19, 21, 22
Large Language Models (LLMs), 6–8, 42, 45, 46, 91, 118, 130, 132, 147, 165, 201
Legacy systems, 249
Lewin's three-stage model, 272
LLMs, *see* Large Language Models (LLMs)
Loan approvals, 308
    address algorithmic bias, 310
    algorithmic bias, 309
    impact and consequences, 309
    root causes of biases, 309
Long-term financial planning
    AI's impact, 173
    decision-making in wealth management, 173
    focus on value-added services, 174
    predict future financial outcomes, 173
    regulatory considerations, 175
    regulatory framework, 175
    strategies for navigating challenges, 174
    technical and ethical challenges, 174
    traditional client-advisor relationship, 174

M

Machine learning (ML), 58, 81, 131, 134, 167, 171, 211
Maersk, 155
Market agility, 68
Market dynamics, 15
Market volatility, 64, 69, 75, 77
Markov chain, 26
Mastercard, 135, 310
McKinsey's 7-S Framework, 272
Microfinance platforms, 139
Microsoft's Azure AI, 97, 100
Mind maps, 258
Mint app, 145
Mitigating bias, 36
Mitigation strategies, 317, 320
ML, *see* Machine learning (ML)
MLOps streamline AI model, 270
Mobile banking, 57, 58, 80
Model drift, 270, 271
Model inversion attacks, 316, 320
Model training, 112
Modern data analytics systems, 89
Modernization, 87
Modern Portfolio Theory (MPT), 170
Monzo, 145
M-Pesa, 184
MPT, *see* Modern Portfolio Theory (MPT)
MT-Bench scores model, 29
Multimodal generative models, 32
Multinational insurance operations, 231

INDEX

N

Natural language processing (NLP), 25, 126, 167, 168, 171, 175, 178, 179, 196, 199
Neural networks, 3, 41
Neuroscience, 21
NLP, *see* Natural language processing (NLP)
Numerai, 150
NYDFS Cybersecurity Regulation, 230

O

On-demand insurance, 220, 222, 226
    benefits for insurers and policyholders, 223, 224
    consumer behavior and insurance purchasing patterns, 223
    cyber insurance
        cyber risks, 224
        generative AI and LLMs, 224
        market growth and acceptance, 224, 225
        policies, 224
    emerging risks and innovative coverage solutions, 225
    property insurance, 222
    technologies enabling UBI, 223
    travel insurance activated by Geolocation, 222
    UBI models, 223
One-shot prompt, 194
OpenAssistant, 30
Open-source community, 30, 53
Operational agility, 86
Operational efficiency, 108
Organizational adaptation, 248
Organizational challenges
    resistance to AI adoption
        cultural factors, 272
        economic considerations and cost implications, 274
        perceived risks and concerns, 273
        regulatory and compliance issues, 274
        technological challenges and complexity, 273
Organizational scalability, 249

P

Partnership on AI, 37, 154, 186
Pattern recognition, 150
PayPal, 127, 135, 141, 150, 168, 270
Permutable AI, 313
Personal Capital, 145, 153
Personalization, 31, 65, 140, 209–211, 214–218, 223, 284, 285, 287
Personalized health insurance, 215–217, 225
    actuarial modeling, 226
    chronic disease management, 221
    continuous integration, 222
    generative AI and LLMs, 221
    insurance product innovation, 222
    pioneering product innovations, 222
    regulatory approval, 226
    wearable-integrated insurance, 221
    wellness programs, 221
Personalized policies, 207–211, 213, 214
Pilot testing, 14, 79
Pioneering product innovations in insurance
    cyber insurance, 219
    data bias, 218

development, cyber insurance
policies, 219
discrimination, 218
ensuring transparency, 218
Generali, 218
generative AI and LLMs, 219, 220
market growth and acceptance,
219, 220
on-demand and usage-based
insurance, 220
on-demand insurance products, 220
traditional insurance product
landscape, 218
Portfolio management techniques, 130,
140, 147, 165
asset allocation models, 148
big data and advanced analytics, 142
challenges and considerations, 142
communication and marketing,
144, 145
customer-centric financial
solutions, 143
customer engagement and AI, 145
customer engagement and loyalty, 145
customer satisfaction, 148
customer service, 144
customized investment strategies, 140
future advancements, 142
HSBC's Amy, 144
optimal asset allocation, 141
risk assessment, 148
strategy development, 147
Swedbank's Nina, 144
Portfolio optimization, 49, 125
Portfolio rebalancing, 170
Positional encoding, 41
Potential disruptions analysis

Decentralized Finance (DeFi), 181
Aave, 181
digital currencies, 181
Bitcoin, 181
economic inclusion, 182
Tala, 182
job displacement and creation, 182
regulatory and ethical
considerations, 182
transformation of traditional Banking
roles, 181
Wells Fargo, 182
Predictive analytics, 72, 90, 104, 106, 108,
110, 141, 149, 167, 178
Professional tone, 187
Progressive's Snapshot program, 212
Prompt engineering, 193, 194
data-driven analysis, 187
define research task, 187
ensure professional tone, 187
example analysis report, VFIAX (*see*
Vanguard 500 Index Fund (VFIAX)
few-shot prompt, 195
investor suitability assessment, 188
key areas of focus, 188
one-shot prompt, 194
provide clear recommendations, 188
RCI technique prompt, 195
real-world example, 187
researcher role/persona, 190
research task, 190
scientific context, 190
set context and objectives, 186
structure report, 187
Psychology, 21
Public perception, 302
Python code, 114, 116

INDEX

Q
Quantum computing, 4, 58, 59, 65, 67, 68, 72–74, 268, 301

R
R&D, *see* Research and development (R&D)
RAG, *see* Retrieval-Augmented Generation (RAG)
RBI, *see* Reserve Bank of India (RBI)
RBI regarding AI integration
    AI implementation and compliance, 290
        accountability for AI decisions, 291
        clarity and transparency, 290
        durability and resilience, 291
        explainability and comprehension, 291
        incorporation, human judgment, 292
        precision and accuracy, 291
        privacy and data protection, 291
        uniformity and consistency, 291
    data bias and robustness, 289
    expectations for financial institutions, 290
    governance, 290
RCI technique prompt, 195, 196
Reactive approach, 87
Regulatory adherence, 13
Regulatory changes, 234
Regulatory compliance, 4, 9, 11, 13, 59, 61, 63, 66, 70, 73, 74, 76, 86, 95, 125, 151, 155, 166, 169, 203, 252, 256, 271, 320
Regulatory foresight and agility, 235
Regulatory frameworks, 53, 129, 175, 177, 231, 268, 289, 290, 292
Regulatory landscape and policy
    ethical guidelines, 37
    explainability, 38
    foster responsible innovation, 38
    GDPR, 37
    guidelines and frameworks, 37
    multimodal and interdisciplinary research, 38
    stakeholders, 38
Regulatory technology (RegTech), 58, 65, 138, 232–234
Reinforcement learning, 28, 51, 55
Reinforcement Learning from Human Feedback (RLHF), 28, 31
Renaissance technologies, 127, 134
Research and development (R&D), 33, 45, 229
Reserve Bank of India (RBI), 289, 290, 292
Responsible AI
    accountability, 19
    balancing innovation with responsibility, 20
    bias, 20
    ethical guidelines, 185
    ethical and societal implications, 20
    inclusivity, 19
    integration of AI, 21
    language intelligence, 21
    neuroscience, 21
    psychology, 21
    sociology, 22
    stakeholder engagement, 185
    transparency, 19, 185
Responsible development and deployment
    AI governance, 299

automation and future of work, 300
banks and financial institutions
    BBVA, 300
    Capital One, 300
    CBA, 300
    HSBC, 300
    JPMorgan Chase, 300
ethical AI design, 299
human oversight, 299
reskilling and upskilling initiatives, 301
Retrieval-Augmented Generation (RAG), 156
    HTTP API, api.py, 159, 160
    configurations, 158, 159
    customizing and scaling, 165
    dependencies installation, 157
    enhancements, 165
    features, 165
    model.py, 161, 162
    Ollama Llama2, 158
    post question, 164
    model.py, 160
    run, 162, 164
Risk management, 49, 59, 61, 63, 69, 71, 75, 77, 79, 81, 105, 124, 131, 141, 147, 207
RLHF, *see* Reinforcement Learning from Human Feedback (RLHF)
Robo-advisors, 131, 136, 147, 168–172, 197, 286, 287
Robotic process automation (RPA), 91, 92
Robust cybersecurity, 12, 285, 288
Robust data governance, 101, 104, 317, 320
Royal Bank of Canada (RBC), 103
RPA, *see* Robotic process automation (RPA)

S

Scalability, 28, 80, 247–249, 253
Sentient Technologies, 149
Sentiment analysis, 48, 50, 125
Siloed sub-systems, 87
Skill development and hiring
    build AI learning culture, 260
    challenges in AI skill development, 262
    collaboration with academia and industry, 262
    core BFSI competencies, 260
    effectiveness, AI skill development programs, 262
    effective recruitment strategies, 260
    key skills, 261
Skill development programs, 262
Sociology, 22
Stakeholder engagement, 76, 185, 264, 298
Strategic integration approach, 247
Stringent data governance practices, 303, 307
Sustainable financial practices
    enhance financial inclusion, 183
        microfinance solutions, 184
        mobile applications, 183, 184
    ESG investing, 183
        Truvalue Labs, 183
    green finance, 183
        ClimateAI, 183
Synthetic data generation, 31

T

Talent acquisition and retention, 60, 62, 64, 67
Talent gap, 260, 262
Technical scalability, 248

INDEX

Technical skills, 261
Technological integration, 61, 63, 66, 68, 70
Technological upgrades, 87
Technology integration, 60
Telematics, 72–74
   in auto insurance, 212
Tokenization, 40
Traditional credit assessment methods, 117
Traditional financial advisory models, 196
Traditional financial advisory services, 168
   AI-driven tools and platforms, 173
   personalized financial advice, 172
   robo-advisors, 168, 169
Traditional fraud detection methods, 118
Traditional health insurance, 221
Traditional risk assessment methods, 203
Training and fine-tuning of models, 51
Transformative potential, 319–322
Transformer architecture, 27, 54
Transformer models, 27
Transparency, 284, 285, 310
Trust, 284, 285, 287
Two Sigma, 149

U

UBS Delta, 148
Usage-based insurance (UBI), 206, 209, 220, 223, 226
U.S. Health Insurance Portability and Accountability Act (HIPAA), 313

V

VAEs, *see* Variational Autoencoders (VAEs)
Vanguard 500 Index Fund (VFIAX), 187
   documentation format, 190–192
   fund performance analysis, 188
   management and strategy evaluation, 189
   peer comparison, 189
   prompt for research, 192
   risk assessment, 189
   suitability for investors, 189
Vanguard Digital Advisor, 173
Variational Autoencoders (VAEs), 26, 33, 34, 55
VFIAX, *see* Vanguard 500 Index Fund (VFIAX)
Virtual assistants, 58, 90, 133, 144, 296, 300, 304
Vitality, 213, 221

W, X, Y

Wealthfront, 127, 132, 136, 141, 143
Wearables
   in health insurance, 213
Wells Fargo, 95, 100, 104, 106, 109, 133, 145, 182
WildBench, 30

Z

Zephyr model, 29, 32
Zest AI, 180

GPSR Compliance

The European Union's (EU) General Product Safety Regulation (GPSR) is a set of rules that requires consumer products to be safe and our obligations to ensure this.

If you have any concerns about our products, you can contact us on

ProductSafety@springernature.com

In case Publisher is established outside the EU, the EU authorized representative is:

Springer Nature Customer Service Center GmbH
Europaplatz 3
69115 Heidelberg, Germany